T0141195

An Ecotopian Lexicon

An Ecotopian Lexicon

Edited by
Matthew Schneider-Mayerson
and Brent Ryan Bellamy

UNIVERSITY OF MINNESOTA PRESS
Minneapolis
London

The University of Minnesota Press gratefully acknowledges the
financial assistance provided for the publication of this book by
a Yale–NUS College Subvention Grant.

Copyright 2019 by the Regents of the University of Minnesota

All rights reserved. No part of this publication may be reproduced,
stored in a retrieval system, or transmitted, in any form or by
any means, electronic, mechanical, photocopying, recording, or
otherwise, without the prior written permission of the publisher.

Published by the University of Minnesota Press
111 Third Avenue South, Suite 290
Minneapolis, MN 55401-2520
http://www.upress.umn.edu

Library of Congress Cataloging-in-Publication Data
Schneider-Mayerson, Matthew, editor. | Bellamy, Brent Ryan, editor.
An ecotopian lexicon / Matthew Schneider-Mayerson and Brent
Ryan Bellamy, editors.
Minneapolis : University of Minnesota Press, [2019] | Includes
bibliographical references.
Identifiers: LCCN 2018058231 (print) | ISBN 978-1-5179-0589-7 (hc) |
ISBN 978-1-5179-0590-3 (pb)
Subjects: LCSH: English language—Foreign words and phrases. |
Environmentalism—Terminology.
Classification: LCC PE1582.A3 E26 2019 (print) |
DDC 422/.403—dc23
LC record available at https://lccn.loc.gov/2018058231

Printed in the United States of America on acid-free paper

The University of Minnesota is an equal-opportunity educator
and employer.

25 24 23 22 21 20   10 9 8 7 6 5 4 3 2

In memory of Beatrice Annie Goranson,

who would have loved this,

and Trout,

who loved seaweed, sunlight, and the wind

*The climate crisis is also a crisis of culture,
and thus of the imagination.*

—AMITAV GHOSH, *THE GREAT DERANGEMENT:
CLIMATE CHANGE AND THE UNTHINKABLE*

To imagine a language is to imagine a form of life.

—LUDWIG WITTGENSTEIN,
PHILOSOPHICAL INVESTIGATIONS

Contents

Another Path

Greetings

Resistance

Dispositions

Perception

Kim Stanley Robinson

Languages change over time when people use new words. Fairly often these new words are borrowed from other languages, and linguists call these *loanwords,* though the return on the loan is never reciprocal, being no more than some kind of social-psychic tribute to the ingenuity of the host language's culture. Probably there should be a word too for this mysterious return on the loan.

Science fiction, consisting of stories set in the future, has often led its writers onto the perilous ground of making up new words, or even new languages. These languages suggest the historical and technological changes that have occurred between the present and the time of the science fiction story, and more generally they create the feeling of estrangement appropriate to science fiction's transport by mental travel to distant times and places. Orwell's controlocracy in *1984* is famously facilitated by Newspeak; Anthony Burgess cleverly portrayed a Russian-inflected English dialect in *A Clockwork Orange*; Russell Hoban conveyed an apocalyptic scenario in the postliterate language of *Riddley Walker*; Suzette Haden Elgin invented a postpatriarchal language in her Native Tongue series. Many other writers have portrayed dense accelerated futures partly by way of a blizzard of neologisms, as in Greg Bear's *Queen of Angels*.

Science fiction writers have also often invented new words, inserted into our familiar English to indicate a new technology or a new mental phenomenon. Two excellent examples are Ursula K. Le Guin's word *ansible,* a communication device that transmits information instantaneously across any distance, and Robert Heinlein's *grok*, which is a kind of telepathic gestalt understanding of some other person or idea. For a decade or more, the popularity of Heinlein's *Stranger in a Strange Land* was such that lots of people used the word *grok* in real life, which is quite an accomplishment for any writer. On the other hand, Heinlein also coined the word *tansstaafl,* from the acronym "There ain't no such thing as a free lunch," and although some libertarians still love to quote this untrue truism, it suggests to me that Heinlein's ability in this realm was hit or miss.

The book you are about to read focuses on individual words or short phrases, either from other languages or invented, that English, and perhaps all languages, could use to better describe our historical situation, which we now often call the Anthropocene—another new word, and one that is very likely to last, defining our period in both human and geological terms, now collapsed to the same thing. One scientist, Eugene Stoemer, invented this word, and another, Paul Crutzen, popularized it. Now it is an important and permanent part of our language and our sense of history. Though this origin story might seem surprising, recall that *ecology* is a word coined by Ernst Haeckel in 1873, while the word *scientist* was coined by William Whewell in 1834. People invent new words! It's obvious when you say it, and yet still a little startling to remember.

Each word or phrase included in this book is more than a word or phrase, being also a concept. Of course all words are concepts, but familiar words contain their concepts in a familiar way, and so seem simpler than new and unfamiliar words. As all the words included here are new to English, they bring new concepts with them, so the short essays accompanying each word are crucial to the success of the project. The essays all move from straightforward definition to discussions of the contexts of the concepts evoked, elucidating larger systems of thought and culture, and casting light on the "long emergency" of the twenty-first century, when anthropogenic climate change will impact the biosphere and all its inhabitants.

So many new words gathered together like this, each bringing with it a new concept and system, creates a dizzying effect. This is good and right, because we live in a dizzying time. What we do now as a global civilization will create one future out of a vast array of possible futures, an array which ranges from utmost disaster to lasting peace and prosperity. The sheer breadth of this range is all by itself extremely confusing, to the point of inducing a kind of mental and emotional gridlock. Anything could happen! So what should we do? Maybe nothing! Maybe we can't do anything!

But we can do things, if we can figure out what they are. Various good futures are achievable, even starting from our current moment of high danger. So some really comprehensive analysis, destranding, and remapping is now part of our necessary work. Inevitably new concepts and new words will emerge—lots of them. So this book's profusion is an accurate foretelling of what will come. It's a kind of science fiction story in the form of a lexicon, and it postulates and helps to create a future culture more articulate and wiser than we are now. Thus by definition it is a utopian science fiction story.

Among other good effects, this book makes us more alert to new words already out there, and it puts us on the lookout for more. Recently I noticed a couple of meteorological phrases new to me, which helped me while I was compressing and revising my Science in the Capital trilogy, written between 2002 and 2006. I saw in my 2016 revision that I had described what we now call an "atmospheric river" a decade before the phrase appeared (at least to me), and I happily inserted it into the text, as being better than earlier names like "pineapple express." Similarly, in the second volume I had described what we now call a "polar vortex," but again, either the term had not yet been invented or it had not yet shifted from the meteorological community to common usage (or to me). Again, I retrojected the phrase into the text.

These kinds of re-visionings are going to keep happening to us all, expanding from small meteorological clarifications such as these I've mentioned to much larger and more important expressions of the zeitgeist, including definitions of actions we can take to wrest history into a good Anthropocene. As these inventions pop in our heads, this delightful lexicon can serve as sourcebook, clarification, diagnostic, and stimulus. It's an already existing example of the way people playing with language can help bring things and events into sharper cognitive focus. Playful and useful: I trust you will enjoy this book, and I hope you will put it to use.

Acknowledgments

While no academic work is truly solitary, edited collections are especially communitarian. This book is the product of a transnational web of scholars, writers, artists, and editors, not to mention the trillions of nonhuman organisms that made our work (and all of human life) possible. Thanks to the nematodes for recycling soil nutrients to enable agriculture; to the trees for producing the oxygen that we breathe every day; and to the bacteria and other microbes that comprise over half of the cells in our bodies. Human friends and colleagues also provided critical support, including Dominic Boyer, Jeffrey Jerome Cohen, Jeff Diamanti, Soo Go, Eva-Lynn Jagoe, Eduardo Lage-Otero, Michael Maniates, Shama Rangwala, Imre Szeman, and Juria Toramae. We thank Chantal Bilodeau for her early contributions. We are grateful to the University of Minnesota Press and our editor, Doug Armato, for supporting this unique project, and we acknowledge the generous assistance of Yale-NUS College, which provided a subvention grant that enabled the inclusion of full-color artwork. We want to express our gratitude to the countless creative thinkers whose work inspired this project, especially Octavia E. Butler, Ursula K. Le Guin, Hayao Miyazaki, Marge Piercy, Kim Stanley Robinson, and Rebecca Solnit. Above all, we thank the authors and artists for their willingness to join us on this journey.

Matthew would like to thank, in particular, Juria for her intellectual, artistic, and emotional support and Trout Fishing in America, Blue, and Maggie for their love and affection. As always, he is grateful for family: Elizabeth Schneider, Anna Schneider-Mayerson, Hal Mayerson, Rebecca Edwards, and Louis Mayerson, among others. This kind of work would not be possible (or fun) without comrades to talk and think with; thank you to Joni Adamson, Jodi Dean, Kei Franklin, Greta Gaard, Jesse Goldstein, Louis Ho, Steffani Jemison, Stephanie LeMenager, Elaine Tyler May, Ella Dawn McGeough, Kevin P. Murphy, David N. Pellow, Daniel J. Philippon, Karen Pinkus, Elizabeth Povinelli, Elise Rasmussen, Alexa Weik von Mossner, Benjamin Wiggins, Audrey Yeo, and Natasha Zaretsky. A residency at the Banff Centre for Arts and Creativity as part of the Banff Research in Culture "Year 2067" program offered an ideal location to share ideas and think about desirable futures. Friends and colleagues at Yale-NUS College were supportive throughout this process, especially Michael Maniates, Marvin Montefrio, Brian McAdoo, Ma Shaoling, Nicole Constable, Joseph F. Alter, Anju Paul, Catherine Sanger, Nick R. Smith, Stuart Earl Strange, Ajay Mathuru, Matthew Stamps, Joanne Roberts, and President Tan Tai Yong. He is indebted to the students in his 2018 "Foundations of Environmental Humanities" and 2019 "Ecotopian Visions" courses at Yale-NUS for their thoughtful discussion of selected chapters. To them, and to you, he says: gyebale!

Brent would not have had the time and space to complete this project without the care, love, and support of Alexandra Carruthers. The companionship of George helped, too, including his constant demand: "Be better!" In the course of editing this book, Brent spoke with many colleagues, comrades, and friends. If you talked and he's missed your name here, hit him up for a copy of the book! Thanks to Adam Carlson, Marija Cetinic, Amy De'Ath, Jonathan Dyck, Danine Farquharson, Veronica Hollinger, David Huebert, Marième Isabel, Rob Jackson, Aaron Kreuter, Lauren Kirshner, Katie Lewandowski, Karla McManus, Tanner Mirrlees, Sean O'Brien, Joseph Ren, and Jill Stanton. Thanks also go to his family (and their animal companions) especially Barbara and Hope, who shared their home in fall 2017, and also to Brent Sr., Wendy, Rebecca, William, Josten, Blue, Max, and Lily; Leigh Ann and Pearl; and Clyde and Lindsay. Thanks are also due to Adrienne for the vital work of helping to bring new hope to the world.

Loanwords to Live With

Matthew Schneider-Mayerson and Brent Ryan Bellamy

For the past decade, dark clouds have been massing on the psychic horizon. Against the trickle of encouraging news about the growth of renewable energy comes a countervailing torrent of numbingly ominous climatic developments. The four elements seem to be assembled like the four horsemen. Air: the world chokes. In May 2019, the concentration of atmospheric carbon dioxide passed 414 parts per million, the highest concentration in millions of years and far beyond what scientists consider safe for humans.[1] Fire: life swelters and burns. From 2010 to 2016, seven successive years registered as having the highest average global temperature on record.[2] Calamitous wildfires have become more common, from Australia to Greenland and everywhere in between. Water: the land floods. In July 2017, an ice shelf twice the size of Luxembourg broke off the Antarctic ice shelf, floating toward warmer waters; a month later, Houston and Mumbai endured devastating floods. Earth: the soil bakes. A drought that started in 1998 in the eastern Mediterranean is likely the most severe of the past nine hundred years, significantly contributing to the mass migration out of Syria.[3]

Under these stresses, the web of life is coming undone. According to many scientists, we have now entered the sixth mass extinction in our planet's history. Some species will adapt; others, tragically, will not. Yet readers of this book do not need more statistics about the climate catastrophe; those of us paying attention are already familiar with the projections. Learning the basic facts about our warming world is only the beginning of reckoning with what needs to be done to mitigate the worst impacts on the planet, nonhuman species, and ourselves. We all struggle to understand how to respond, emotionally and politically, collectively and as individuals, to the advent of the Anthropocene, the current geologic epoch in which humanity acts as the dominant world-shaping force.[4] There is no doubt that the Earth will be profoundly transformed in the coming decades. The situation is dire, but instead of catastrophism, this book offers something else: conceptual tools to help us imagine how to adapt and flourish in the face of socioecological adversity.

To forge such tools, we need to take stock of where we are. We need to accept that what we might have imagined to be stable, consistent, and normative ceaselessly shifts underfoot. As philosopher Allen Thompson argues, what many of us in the late twentieth and early twenty-first centuries saw as "normal" will likely be seen, "in the full course of human history," as "an outstanding aberration."[5] As the scale and fallout of climate change, ocean acidification, mass extinction, and other processes become increasingly undeniable and unavoidable, we will need to change our cognitive maps of the world. What are the psychological analogs of calving icebergs and drowned coastlines? What new cultural constructions will arise along with renewable infrastructures? Though we cannot accurately predict the future, it is now reasonable to "expect that many, but not all, of the culturally embedded perspectives and habits basic to late twentieth century life will not continue."[6] Indeed, the question is not whether the world will change, but how. Through the medium of language, this book presents possibilities for the cultural maps of better futures.

WHAT IS TO BE IMAGINED?

Over the last two decades, one of the most popular rallying cries from radical activists and scholars in the West has been "another world is possible." It has been chanted at thousands of protests, scribbled on countless concrete borders worldwide, and repeated over 800,000 times on the Internet (according to a recent web search). "Another world is possible" is a conscious rejoinder to the assertion that "there is no alternative" to the status quo, unjust though it may be. Traced to nineteenth-century liberal political theorist Herbert Spencer, "there is no alternative" was popularized by British prime minister and grand wizard of austerity Margaret Thatcher, who used it so frequently to justify her callous deregulation and market fundamentalism that it acquired its own acronym: TINA. Thatcher was reading from an authoritarian playbook: fascists and totalitarians understood that by restricting the imagination and consideration of alternative possibilities—of politics, policy, and social life—citizens would resign themselves to the order of things, thereby enabling further manipulation by political and financial elites. In the face of catastrophic ecological collapse, novel versions of "there is no alternative" are intoned today, by conservatives recklessly dedicated to fossil-fueled capitalism as well as by comfortable liberals who wish to see only gradual, incremental, and nonthreatening responses to an increasingly turbulent status quo.

"Another world is possible" is a worthy maxim, but without further elaboration it stands like a narrow bridge to a destination shrouded in mist. Would you want to journey there? When radical activists and scholars describe what lies on the other side, they often write manifestoes or lists of principles, yet these dry or polemical formulations rarely stir the passions. This is where speculative fiction holds great promise. Though the mist always clings to visions of different worlds, the very act of imagining the future enables a radical departure from the trajectory of the present.[7] As "barely audible messages from a future that may never come into being," speculative visions offer critical perspectives on the present and remind us of the infinite possibilities of life on Earth, as well as the ways that we might bring some of them into being.[8]

When a wave of fiction explicitly concerned with climate change arrived a decade ago, it held incredible potential as a kind of literary time machine, transporting us to the foreign country of the distant future. Unfortunately the genre has not yet realized that promise, as many authors have recycled familiar apocalyptic tropes that describe diminished, destroyed, and denatured worlds.[9] Though packaged as cautionary tales that might spark us to take action, social scientists report that dystopic visions tend to do the exact opposite.[10] In the few cases where authors have constructed desirable environmental futures, they often portray their inhabitants with familiar subjectivities: people of the future, they're just like us! Consider the characters of Kim Stanley Robinson's compelling 2017 novel, *New York 2140*, who are comfortably recognizable: social media celebrities, greedy financial traders, and treasure-seeking orphans.[11] This representational tendency exists so that readers have something familiar to hold on to while being transported to an unfamiliar world, but it can cast a misleading spell. Historians of social life, emotion, and the everyday report that our ancestors were not just like us; there is little reason to believe that our descendants will be any less different. Acknowledging the probable alterity of future inhabitants of Earth is a first step in meeting the obligation to change our culture, our selves. This imperative is one of the goals of this book, and the reason its authors draw on speculative fiction, anthropology, and the sociology of existing subcultures: these fields help us remember that other worlds already exist, while history reminds us that change is inevitable.

We now know that even our sensual apperception is not "natural" but rather dependent on fundamental, unconscious, and culturally specific worldviews, norms, and scripts. However, this knowledge alone—like the phrase "another world is possible"—does not help us imagine, cognitively or emotionally, better worlds. Especially at this historical moment, imagination is not frivolous but crucial: as cognitive psychology's simulation heuristic argues, people view a possible event as plausible to the extent that they can imagine it.[12] Small tweaks to the status quo—such as replacing fossil fuel consumption with centralized and corporate-controlled solar and wind power—are much easier to imagine than wholesale transformations. This perceptual tendency has profound consequences. Even the most felicitous future is not worth struggling for if it is not considered "realistic," if it seems like our hopes cannot be realized. Because of this poverty of imagination, we are left fighting for minor shifts in policy that most scientists see as wholly insufficient, or we find ourselves faithfully trudging toward a destination we cannot even envision. As ecofuturist Alex Steffen notes, "It's literally true that we can't build what we can't imagine. . . . The fact that we haven't compellingly imagined a thriving, dynamic, sustainable world is a major reason we don't already live in one."[13]

Enter the volume you hold in your hands. It does not pretend to offer a fully fleshed-out description of an ecotopian future, let alone a road map to get there. What it does present is an assortment of conceptual tools and a prismatic window into the ecological multiverse. In the tradition of speculative fiction authors such as Octavia E. Butler, Ursula K. Le Guin, Marge Piercy, and Kim Stanley Robinson, it presents thirty terms and concepts to jump-start the critical process of imagining and eventually realizing better futures. Unlike most speculative fiction, however, these visions revolve around describing different ways of inhabiting this fractured planet to which we are inextricably bound. As Jacques Mesrine notes, "There is no other world. There's just another way to live."[14] In that spirit, you might read these entries not only as descriptions of unfamiliar ideas but as fun house mirrors in which you can glimpse yourself as radically otherwise.

LANGUAGE AND REALITY

What role could language play in a Great Transition? As ecotheologian Thomas Berry asserts, "Our challenge is to create a new language, a new sense of what it is to be human. . . . This brings about a completely new sense of reality and of value."[15] There is little question that language reflects the material and conceptual worlds that we inhabit—consider, for example, the copious critters and place-names that are slowly being edged out of dictionaries and common usage in favor of recent coinages such as *browsability* and *post-truth*. Classical thinkers from Aristotle to Giambattista Vico have suggested that this is a two-way street: language might also have a powerful influence on perception and thought. In the early twentieth century, anthropologists began to focus their attention on this hypothesis, now known as linguistic relativity. One of its leading theorists, Benjamin Lee Whorf, summarizes this notion in a tellingly ecoimperialist metaphor:

> We dissect nature along lines laid down by our native languages. . . . The world is presented in a kaleidoscopic flux of impressions which has to be organized by our minds— and this means largely by the linguistic system in our minds. We cut nature up, organize it into concepts, and ascribe significance as we do, largely because we are parties to an agreement to organize it in this way—an agreement that holds throughout our speech community and is codified in the patterns of our language.[16]

The "thick" version of this idea, commonly referred to as the Sapir–Whorf hypothesis, is linguistic determinism: the claim that language not only influences but determines and therefore constrains perception and thought. Though repudiated by contemporary social scientists, linguistic determinism has been part of the motivation for the intentional construction of over nine hundred artificial languages, from Lingua Ignota (conceived by twelfth-century German mystic Hildegard von Bingen) to Loglan (developed in the 1950s to transform its speakers into logical thinkers) to Láadan (a feminist language created in the 1980s).[17] The 2016 film *Arrival* stretches the Sapir–Whorf hypothesis to the point of caricature: learning an alien language with no linear concept of time somehow enables its speakers to see the future.

Is there an ideal ecological language that would instantaneously usher in a true culture of sustainability? Unfortunately, no. The labor of constructing a better future will take many forms, though the power of language and concepts should not be understated. Linguistic relativity has often focused on the influence of grammatical structures, yet "words wield tremendous influence over human thought and action" too, notes linguistic anthropologist Sean O'Neill.[18] Who hasn't had a memorable moment of recognition at learning a new term that encapsulates and thereby crystallizes an existing idea or feeling? The power of novel terms to establish practices as normative might be demonstrated by the recent neologisms *selfie* and *binge-watch*: what had been embarrassing can become, once named, communicable and commonplace.

Many environmental thinkers agree that language should be one site of analysis and intervention as we confront the Anthropocene. Scholars of environmental studies and the environmental humanities have published thoughtful critiques of familiar but problematic terminology (such as "nature," "culture," and "the environment") and highlighted newly resonant terms (including "denial," "extinction," and "apocalypse").[19] British author Robert Macfarlane has published a loving lexicon of reenchantment via local place-names and called for a "desecration phrasebook" of our near future, full of neologisms for trash vortices, geoengineering drizzles, and oil-soaked tidemarks.[20] Artists such as the Bureau of Linguistical Reality[21] have crowd-sourced the coinage of Anthropocene terminology, leading to suggestive terms such as *ennuipocalypse* (a slow-motion collapse) and *shadow time* (a feeling of living simultaneously in two different temporal scales). We are fellow travelers with these thinkers and artists, but this book offers a slightly different perspective. Our authors present their loanwords as conceptual tools for reckoning with the environmental, political, social, and philosophical challenges of the Anthropocene, today and in the decades to come. This lexicon does not stop at critique of what exists today; it argues for what could or even should be. To think of language in this way is to implicate the daily choices we make as individuals and communities—to utter one word instead of all others is to shape the direction of our living language, consciously or not. Insofar as every choice shapes the cognitive frames we inhabit, our future is established not only through dramatic historical events but also through gradual accretion: moment by moment, act by act, word by word. In this spirit, this lexicon does not merely describe unfolding disasters but offers buoyant linguistic and conceptual tools for the collective construction of a future that is more just, equitable, pleasurable, and truly sustainable for *Homo sapiens* and the millions of species with whom we gratefully share this planet.

CONCEPTS IN MIGRATION

These suggested additions to the current global lingua franca known as Standard American English do not come to us from the ether. They hail from speculative fiction, subcultures of resistance, and other languages and cultures. This last category is particularly important. While resurrecting what has been forgotten and constructing new things ex nihilo will be necessary, we must seek appropriate responses to our collective challenges with the full tapestry of human experience in mind. Whatever may come, humility is a surefire virtue in the Anthropocene, and welcoming new and old friends with different experiences and forms of knowledge will be especially important in this era of change, turbulence, and migration. We acknowledge that such borrowing, however conscientious, carries the risk of cultural appropriation, especially because English speakers have been and continue to be responsible for staggering material and cultural theft from Indigenous peoples and people of color. With this history firmly in mind, we believe it is crucial to learn from and think with other cultures and subcultures in these perilous years, and that this can be achieved without the symbolic violence of romanticizing the outlooks or reducing the complexity of previously unfamiliar worlds and worldviews. As environmental critic

Ursula Heise has noted, "The environmentalist ambition is to think globally, but doing so in terms of a single language," or a single culture, "is inconceivable."[22]

In this sense, the loanword is a fruitfully relational linguistic category. Loanwords are terms that are adopted into one language from another without translation.[23] Their irregular spelling and pronunciation thus advertises their difference, demonstrating that language, like culture, is always heterogeneous and historied. Contemporary English is unimaginable without loanwords, from *government* (French) to *chocolate* (Nahuatl) to *ecstasy* (Greek). Indeed, English is a particularly eclectic language; according to the World Loanword Database,[24] 42 percent of English words are loanwords, compared to only 2 percent for Mandarin Chinese. While loanwords are not truly loaned in the sense that they are incorporated into one language without the prospect of being returned later, they certainly constitute a gift. Indeed, linguists sometimes refer to their language of origin as the "donor language." While a gift can be seen as a form of theft (i.e., uncompensated incorporation), it might also be viewed in the context of the gift exchanges that occur in many cultures. Anthropologists attest that gift giving is not a simple or isolated act but rather serves to weave communities together into dense networks of mutual indebtedness, exchange, and interdependence. Accepting these loanwords, then, makes English speakers obligated to return the favor with gratitude, respect, and equal moral consideration.

Because this book is written in English, it is focused on the value that these loanwords might bring to English speakers. Given English's global popularity—in 2017 it was estimated that there are 1.12 billion English speakers worldwide—and the disproportionate and continuing responsibility of many English speakers for global environmental injustices, we see this as an important intervention. However, English is not the language of the majority of the (human) world. Mandarin is spoken by 1.1 billion people, Hindustani by 697 million, Spanish by

512 million, and Arabic by 422 million. Mandarin and Arabic are the native tongues of far more people worldwide than English—909 million and 442 million, respectively, to 378 million.[25] None of the editors, authors, or artists involved in assembling this ecotopian lexicon believe that English, with or without these loanwords, ought to replace any of the seven thousand languages that are spoken today. Indeed, linguistic imperialism is often a means of cultural expansionism, which reinforces the notion that dominant cultures are more advanced and desirable.[26] Rather than propose a linguistic monocrop, we hope that these loanwords might highlight the world's linguistic and cultural diversity, expand the collective imagination of environmental possibilities, and even inspire similar projects.

Words can only take us so far, of course. It is of little surprise that when it comes to the imagination, we almost always rely on visual metaphors. Sight enjoys a remarkable dominance over the other senses—how often do people long to taste, touch, or hear the future? To add an additional imaginative (and mnemonically sticky) layer to this project, we challenged fourteen artists to respond to selected entries with original artwork, along with a short statement reflecting on their process and product. These artists hail from eleven countries and draw on a wide range of visual styles and artistic traditions, from Buddhist sculpture to sgraffito black-and-white etchings to Día de los Muertos prints. The result is a transmedia conversation between the originary author, culture, or subculture; a critical thinker; and an artist. These images can be found in the color plate in the center of the book; they are also available online for purchase as T-shirts, tote bags, handkerchiefs, and stickers (www.ecotopianlexicon.com). All proceeds from these items and this book will go toward a fund to support creative political and cultural interventions focused on addressing climate injustice—the disproportionate vulnerability and suffering from the consequences of climate change by those who are least responsible.

LOANWORDS TO LIVE WITH: AN ECOTOPIAN LEXICON

We asked each of our contributors to compose an entry describing a word or phrase that would aid us in our collective task of living well in the Anthropocene. We requested terms whose adoption might accomplish much-needed psychological, social, cultural, and political work, and we encouraged contributors to exercise the creativity that this moment demands. Each entry would need to accomplish three basic tasks: introduce the term in its original context; identify its ecological, ecopsychological, ecosocial, or ecopolitical potential; and describe how it might be applied in common usage in English.

These minimal instructions produced a grounded yet vertiginous collection of essays. We present these focused meditations alphabetically—the traditional organization of a book of keywords. But unlike those works, which generally offer a familiar list, our table of contents will be, for most readers, dizzying and even intimidating. That is very much the point: to think we already know the range of social, cultural, phenomenological, psychological, political, and economic options for responding to the existential challenge that we face is to foreclose the possibility of something radically different, something better. (For readers that prefer some guidance, we offer Another Path, which you can take by turning to the loanword listed at the end of each entry.) We hope you take the strangeness of these terms as an invitation to join us on a critical and imaginative journey. We also hope that you will share with others the sense of wonder and possibility that this book intends to animate. Give *An Ecotopian Lexicon* as a holiday gift. Bring it to parties. Leave it in a coffee shop, waiting room, or library for a stranger to find.

To make the most of this historical moment requires that all concerned, creative, and thoughtful people—including you, dear reader—play a role. Culture is ultimately a fragile tapestry that we weave together. As the dominant culture is worn thin by the lashing cataclysms of the Anthropocene, why not choose to weave differently? In

Jorge Luis Borges's short story "Tlön, Uqbar, Orbis Tertius," artifacts from an invented world, Tlön, begin to appear. This fictional universe, it turns out, was not the fabrication of a lone genius but a diverse range of actors, including "astronomers, biologists, engineers, metaphysicians, poets, chemists, algebraists, moralists, painters, geometers"—just the sort of motley coalition that we need today.[27] From a solitary mention in a single dictionary to a few coins, the imaginative vision of a radically different world slowly supersedes reality. People simply prefer to live in Tlön, and so they do, transforming their ideology, language, poetry, and even history. Borges's story illustrates the extent to which we might exert collective control over the perceptual, cultural, social, and political worlds that we choose to inhabit. Consider each of these loanwords, then, as a Tlönian seed buried in the backyard of your mind, waiting to be watered. Let's see what grows.

NOTES

1. "Carbon Dioxide Levels Hit Record Peak in May," NOAA Research News, June 4, 2019, https://www.research.noaa.gov/.
2. "Vital Signs: Global Temperature," NASA Global Climate Change, https://climate.nasa.gov/. These records date back to 1880.
3. Ellen Gray, "Drought in Eastern Mediterranean Worst of Past 900 Years," NASA Global Climate Change, February 29, 2016, https://climate.nasa.gov/; Colin P. Kelley, Shahrzad Mohtadi, Mark A. Cane, Richard Seager, and Yochanan Kushnir, "Climate Change in the Fertile Crescent and Implications of the Recent Syrian Drought," *Proceedings of the National Academy of Sciences of the United States of America* 112, no. 11 (2015): 3241–46.
4. Jan Zalasiewicz, Mark Williams, Alan Smith, et al., "Are We Now Living in the Anthropocene?," *GSA Today* 18, no. 2 (2008): 4–8. For a full description of the history and implications of the Anthropocene, see Ian Angus, *Facing the Anthropocene: Fossil Capitalism and the Crisis of the Earth System* (New York: Monthly Review Press, 2016).
5. Allen Thompson, "Radical Hope for Living Well in a Warmer World," *Journal of Agricultural and Environmental Ethics* 23, no. 1–2 (2010): 45.
6. Thompson, "Radical Hope," 45.
7. For the relationship between environmentalism and utopianism, see David Pepper, "Utopianism and Environmentalism," *Environmental Politics* 14, no. 1 (2005): 14;

Marius de Geus, *Ecological Utopias: Envisioning the Sustainable Society*, trans. Paul Schwartzman (Utrecht: International Books, 1999); and Ernest Callenbach's classic novel *Ecotopia*, first self-published in 1975.

8. Fredric Jameson, "The Politics of Utopia," *New Left Review* 25 (2004): 54.

9. See Matthew Schneider-Mayerson, "Climate Change Fiction," in *American Literature in Transition, 2000–2010*, edited by Rachel Greenwald Smith (Cambridge: Cambridge University Press, 2017), 309–21.

10. See, for example, Per Espen Stoknes, *What We Think about When We Try Not to Think about Global Warming: Toward a New Psychology of Climate Action* (Chelsea, Vt.: Chelsea Green, 2016); and Matthew Schneider-Mayerson, "The Influence of Climate Fiction: An Empirical Survey of Readers," *Environmental Humanities* 10, no. 2 (2018).

11. In fact, some of Robinson's protagonists are familiar because they are figures from classic (mostly American) literature, from Mark Twain to Samuel Beckett. See Wai Chee Dimock, "5,000 Years of Climate Fiction," Public Books, June 28, 2017, https://www.publicbooks.org/.

12. Daniel Kahneman and Amos Tversky, "The Simulation Heuristic," in *Judgment under Uncertainty: Heuristics and Biases*, ed. Daniel Kahneman, Paul Slovic, and Amos Tversky, 201–8 (Cambridge: Cambridge University Press, 1982).

13. Alex Steffen, quoted in Meir Rinde, "Imagining a Postcarbon Future," Science History Institute, Distillations, fall 2016, https://www.sciencehistory.org/.

14. Jacques Mesrine, quoted in the Invisible Committee, *To Our Friends* (South Pasadena: Semiotext(e), 2015), 9.

15. Thomas Berry, "The Ecological Age," in *The Dream of the Earth* (San Francisco: Sierra Club Books, 1988), 42.

16. Benjamin Lee Whorf, *Language, Thought, and Reality: Selected Writings of Benjamin Lee Whorf* (Cambridge, Mass.: MIT Press, 1956), 213–14.

17. See Arika Okrent, *In the Land of Invented Languages: Adventures in Linguistic Creativity, Madness, and Genius* (New York: Spiegel & Grau, 2010).

18. Sean O'Neill, "Mythic and Poetic Dimensions of Speech in Northwestern California: From Cultural Vocabulary to Linguistic Relativity," *Anthropological Linguistics* 48, no. 4 (2006): 309.

19. See, for example, Joni Adamson, William A. Gleason, and David N. Pellow, eds., *Keywords for Environmental Studies* (New York: NYU Press, 2016); Imre Szeman, Jennifer Wenzel, and Patricia Yaeger, eds., *Fueling Culture: 101 Words for Energy and Environment* (New York: Fordham University Press, 2017); Emily O'Gorman and Kate Wright, "Living Lexicon of the Environmental Humanities," http://environmentalhumanities.org/lexicon/; and Cymene Howe and Anand Pandian, "Lexicon for an Anthropocene yet Unseen," Culanth, January 21, 2016, https://culanth.org/.

20. Robert Macfarlane, *Landmarks* (London: Hamish Hamilton, 2015); Robert

Macfarlane, "Desecration Phrasebook: A Litany for the Anthropocene," New Scientist, December 15, 2015, https://www.newscientist.com/.

21. https://bureauoflinguisticalreality.com/.

22. Ursula K. Heise, "The Hitchhiker's Guide to Ecocriticism," *Publication of the Modern Languages Association* 121, no. 2 (2006): 513.

23. For more information on loanwords as a linguistic phenomenon, see Martin Haspelmath and Uri Tadmor, eds., *Loanwords in the World's Languages: A Comparative Handbook* (Berlin: De Gruyter Mouton, 2009).

24. https://wold.clld.org/vocabulary.

25. Gary F. Simons and Charles D. Fennig, eds., *Ethnologue: Languages of the World*, 21st ed. (Dallas, Tex.: SIL International, 2018), http://www.ethnologue.com.

26. See Robert Phillipson, *Linguistic Imperialism* (Oxford: Oxford University Press, 1992), and *Linguistic Imperialism Continued* (New York: Routledge, 2010).

27. Jorge Luis Borges, "Tlön, Uqbar, Orbis Tertius," trans. James E. Irby, in *Labyrinths: Selected Stories and Other Writings*, ed. Donald A. Yates and James E. Irby (New York, 1964), 7–8.

Pronunciation: (Blow a stream of air lightly
across the back of your hand.)

Part of Speech: Various

Provenance: Dolphinese

Example: That UCB comedy sketch about a BP board
meeting struggling to clean up all the coffee spilled at
their table really ~*~ me when I watched it on YouTube.

Imagine a loanword that comes from dolphin speech. Such a "word" would, by necessity, emerge within the dolphin's oceanic milieu, a salty-smooth volume of lightened gravity. The word would reflect the dolphin's own sensory capabilities (like echolocation) and cognitive predispositions—perhaps a particular whistle or sequence of clicks. Yet upon translation into English, the word would leave behind both the oceanic and Delphic bodies of its originary formation. Indeed, if I dredge this word up from the sea, dripping and disoriented, would it even make sense to you, reader, whom I imagine perusing this text in the dry comfort of a chair? Would too much be lost upon the removal of this word from the watery milieu of dolphin sociality, imported into the contexts of human terrestrial habitation? Wouldn't the translational— or better, transductive—process of making the aquatic word sensible in English be destined for failure?[1]

As Friedrich Kittler wrote over a quarter century ago, "To transfer messages from one medium to another always involves reshaping them to conform to new standards and materials."[2] Yet if one's goal is not mimesis or exact similitude—the fantasy of perfect translation—then perhaps there is something to learn from the attempt to imaginatively translate a "word" from dolphin to English. Language is never exact or fully literal, anyway; take for example metaphor (e.g., the ocean of the unconscious), which involves shuttling between two different figures, a matter of transport. Michel de Certeau recounts that in modern Athens, public transportation vehicles are called *metaphorai*. Thus, "To go to work or come home, one takes a 'metaphor'—a bus or a train. Stories could also take this noble name: every day, they traverse and organize places; they select and link them together; they make sentences and itineraries out of them."[3] If metaphors shuttle between places, then perhaps it is as metaphors that we might persist in imagining a loan-word from dolphinese. Perhaps a loanword from dolphin might be useful and expressive in contexts that we already figure as aquatic—like media, with its informational flows and floods, its content streaming, and its web surfing. This is the speculative challenge that I take up, with help from marine mammologist Denise Herzing.

In a 2013 TED talk, Herzing recounts the variety of body postures and phonations that dolphins use to communicate—behaviors that she has observed over the course of her twenty years studying Atlantic dolphins in the Bahamas. The two best-studied dolphin sounds are "signature whistles" (which function socially, like a name) and "echolocation clicks" (which function technically, to sense the immediate environs). Dolphins can also "tightly pack these clicks together and use them socially. For example, males will stimulate a female during a courtship chase." In the talk, Herzing pauses conspiratorially and adds, "You know, I've been buzzed in the water. [laughter] Don't tell anyone, it's a secret. And you can really feel the sound, that's my point." Indeed, "sound can actually can be felt in the water because the acoustic impedance of tissue and water is about the same, so dolphins can 'buzz' and *tickle each other at a distance*."[4]

I choose the Delphic buzz of "tickling at a distance" as my loan "word," a noise that stimulates from afar, a vibratory jouissance borrowed from entities that spend their lives in the ocean. I use the notation "~*~" for its visual similarity to sound waves and ocean waves (~), punctuated by (*) to denote the bodies that exists within/through them. Recent studies confirm what many of us already know: you can't tickle yourself. Rather, it is the unexpected surprise of touch that causes new vibrations of pleasure and discomfort.[5] Thus, the fact that ~*~ has to do with the heard/felt sensation of tickling implies the presence of at least one other being. As I explore here, the verbal/haptic behavior of "tickling at a distance"—or, for us, the synesthetic experience of "feeling sound"—might generatively enter the English lexicon as a way of talking about the capacity of media—wave media, fiber-optic media, all kinds of vibratory transmissions—to affect humans from a distance. As an environment for thought, the ocean opens the door for a new kind of comparative studies between media and the thalassic wild.

However, I am not the first media theorist to wryly turn to dolphins. John Durham Peters develops an expanded definition of "media" with dolphins as his guides, imagining all kinds of phenomena that count as media for dolphins—yet one that avoids instances of dolphin aggression or promiscuity, regarding dolphins as "medieval theologians did angels" in a literal way—as innocents—in addition to being "entities helpful for thought experiments about intelligence in different media."[6] Although, like Peters, I began with what scientists know about dolphins—Herzing's observation that dolphins tickle each other at a distance with sound—my goal is not to provide a more accurate portal into what it is like to be a dolphin. I am more interested in the ways that ~*~ transports us from ocean to terrestrial media and back again.

One of the challenges in imagining a pronunciation for ~*~ is that sound propagates differently underwater than in air. If ~*~ means to "feel sound and be tickled by it"—something that we don't normally think about, except at loud concerts—then what kind of pronunciation

could approximate how ~*~ sounds/feels underwater? I imagine ~*~
being pronounced by lightly blowing a stream of air across the sensitive back of one's hand, a gesture that might make you shiver. In this
way, the pronunciation of ~*~ suggests the feeling of being tickled at a
distance by someone/something not touching you with their body, an
activity without an existing etymological precedent. Indeed, we often
lack the language adequate to physical experiences that we have no
empirical basis for, like alien languages or breathing through gills. Perhaps Octavia E. Butler puts it best when she has an alien in her novel
Dawn try to explain its feeling of grief to a human being: "Move the
sixteenth finger of your left strength hand."[7] Because we lack the etymological precedent for a word relating to the experience of being tickled
at a distance, ~*~ is of uncertain grammatical status. Is it a noun or
concept, a verb, a preposition, or an entire sentence? I am reminded of
Ludwig Wittgenstein's discussion of whether the command "Slab!" is a
word or an entire sentence. "If a word," he writes, "surely it has not the
same meaning as the like-sounding word of our ordinary language, for
in (2) it is a call. But if a sentence . . . perhaps it could be appropriately
called a 'degenerate sentence.' . . . But why should I not on contrary
have called the sentence 'Bring me a slab' a lengthening of the sentence
'Slab!'?"[8] Although Wittgenstein's examples in *Philosophical Investigations* (1953) are consistently terrestrial (after all, as he famously wrote, to
imagine a language is to imagine a form of life), they are still instructive
for thinking about questions of dolphin speech. Like "Slab!" ~*~ could
be either a word or a sentence. ~*~ both names and performs the phenomena it names all at the same time, an example of what J. L. Austin
called "illocutionary" words.[9] Alternatively, it may even be an example
of onomatopoeia, the formation of a word that resembles or evokes the
sound that it describes.

Initially, I imagined ~*~ like cat's meow, something one might
invoke in a variety of situations or grammatical positions to express a
command (Meow! Pay attention!), a question (Meow? Are you OK?), a

state of being (Meow. I'm frustrated), and the like. In this sense, ~*~ would be not unlike texting an emoticon, a sign that can be embedded in a variety of context and has no given pronunciation. However, Gregory Bateson—who also uses cats meowing to think about dolphin phonation—takes a different course of argument. Bateson references the cat's meow as an example of a communication that is not about "things" per se but about relationships:

> If we were to translate the cat's message into words, it would not be correct to say that she is crying "Milk!" Rather, she is saying something like "Ma-ma!" Or, perhaps still more correctly, we should say that she is asserting "Dependency! Dependency!" The cat talks in terms of patterns and contingencies of relationship, and from this talk it is up to you to take *a deductive* step, guessing that it is milk that the cat wants.[10]

At first glance, Bateson's passage presumes a kind of species hierarchy (not uncommon in studies of animal communication) by characterizing the cat's meow as a kind of primitive and simplistic communication.[11] By introducing the example of the cat's meow, one would guess that dolphin vocalizations could also be primarily about relationships rather than linguistic signs that refer to things outside themselves, and thus may also be seen as somewhat primitive. Could the clue to decoding dolphin come from a better understanding of the relationship happening at particular moments of vocalization? Bateson takes a different course, noting the way we can sense the emotiveness of "meow" but cannot guess at the emotive quality of dolphin phonations:

> We terrestrial mammals are familiar with paralinguistic communication; we use it ourselves in grunts and groans, laughter and sobbing, modulations of breath while speaking,

~*~

and so on. Therefore we do not find the paralinguistic sounds of other mammals totally opaque. We learn rather easily to recognize in them certain kinds of greeting, pathos, rage, persuasion, and territoriality, though our guesses may often be wrong. But when we hear the sounds of dolphins we cannot even guess at their significance.[12]

What strikes Bateson, then, is the opacity of dolphin speech—that one can't even guess at its emotional content; we have no idea what they are saying, even though we can usually guess (based on paralinguistic cues) what another human being speaking a foreign tongue means. Dolphin speech seems to be of a different order than the cat's meow.

Yet because this loanword entry is a speculative fiction, my goal is not to decipher dolphinese but rather to imagine how an underwater vibration intended to tickle (pleasurably or uncomfortably) could serve as a metaphor in terrestrial/human contexts. To return to my hypothesis—that a loanword from dolphin might be useful and expressive in contexts that are full of aquatic figurative language, such as the informational flows of media and web surfing—it may help to generate a list of situations in which "tickling at a distance," or ~*~, might be of use. Consider your phone vibrating in your pocket when you get a text, a light tickling sensation that alerts you to an as yet unopened message. Maybe the text is a GIF, a video that has been roughly looped on repeat to elicit a laugh, that involuntary movement of your diaphragm muscles, a kind of full-body vibration. The Apple Watch even allows you to text your heartbeat. Think of radio, podcasts, music—vibrations mediated by airwaves and cables that cause vibrations traveling over a long distance to resonate in your eardrum. Or think about a video game controller that vibrates in your sweaty hands when your avatar takes on damage. Because dolphins are promiscuous, think about the existence of remotely controlled sex toys, operated by a partner from across the

Internet. If we take literally the metaphor that "information is water," then we might view each of these media as types of propagating channels that carry vibrations over a distance to affect a receiver. Alternately, ~*~ might also operate across global scales, indicating the vibrations of an earthquake, the force of a tsunami, or the rush of monsoon winds—phenomena that transcend individual senses. Indeed, ~*~ might best be used to communicate ecological affects experienced over a distance, evoking synesthetic registers of feeling/hearing.

NOTES

1. See Stefan Helmreich, "Transduction," *Keywords in Sound*, ed. David Novak and Matt Sakakeeny (Durham, N.C.: Duke University Press, 2015).
2. Friedrich Kittler, *Discourse Networks 1800/1900* (Palo Alto, Calif.: Stanford University Press, 1990), 265.
3. Michel de Certeau, *The Practice of Everyday Life* (Los Angeles: University of California Press, 1984), 115.
4. Denise Herzing, "Could We Speak the Language of Dolphins?," TED Talk, February 2013, https://www.ted.com/talks/denise_herzing_could_we_speak_the_language_of_dolphins. My emphasis.
5. See Sarah-Jayne Blakemore, Daniel Wolpert, and Chris Frith, "Why Can't You Tickle Yourself?," *Neuroreport* 11, no. 11 (2000): R11–16.
6. John Durham Peters, *The Marvelous Clouds: Toward a Philosophy of Elemental Media* (Chicago: University of Chicago Press, 2015), 56.
7. Octavia E. Butler, *Lilith's Brood* (1987; reprint, New York: Grand Central Publishing, 2000), 225.
8. Ludwig Wittgenstein, *Philosophical Investigations* (London: Blackwell, 1953), § 19.
9. J. L. Austin, *How to Do Things with Words* (Cambridge, Mass.: Harvard University Press, 1975).
10. Gregory Bateson, *Steps to an Ecology of Mind* (Chicago: University of Chicago Press, 1972), 372.
11. Of course, one could imagine other situations where the range of emotions in "meow" is much more complex, such as instances of grief.
12. Bateson, *Steps*, 376.

Apocalypso

Sam Solnick

Pronunciation: a-pock-a-lip-so
(ə'pɑkə‚lɪpsoʊ)

Part of Speech: Noun

Provenance: Contemporary poetry
(Evelyn Reilly's *Apocalypso*)

Example: Apocalypses tell you that all is lost because
the world is about to end with either a whimper or a bang.
Apocalypsos show us that although the situation may
look (really) bad, you should not give up, because while
some things are coming to an end, others are being born.
So stop whimpering; start dancing.

"Two thousand zero zero party over, oops, out of time," sings Prince
in his millenarian Cold War classic, "1999." The song bears several
hallmarks of apocalyptic texts. Like the biblical apocalypses of Daniel
or John of Patmos, Prince presents a prophetic vision ("I was dreaming when I wrote this").[1] It also belongs in the tradition of nuclear and
environmental apocalypses emerging from what literary critic Frederick Buell describes as a "world-historical change in humanity's position" when, for the first time, advancements in humans' technologies
gave us the power to initiate our own ending on a global scale.[2] In the
nuclear flash, we become destroyers of worlds. Accordingly, for Prince,
"war is all around us" and "everyone's got a bomb." Armageddon looms

in our minds, enveloping us, and when the end is nigh, you might as well party hard. Hence, despite its jubilant chorus, the song reproduces some of the more problematic tendencies of apocalyptic art, including pessimistic nihilism ("Tryin' to run from the destruction, you know I didn't even care") as well as an exclusion of those who do not want to revel in revelation ("If you didn't come to party, don't bother knockin' on my door").[3]

We frequently use the word *apocalypse* as shorthand for cataclysm or destruction on a grand scale, but the term comes from the Greek ἀποκάλυψις, meaning "un-covering" or "disclosure." The etymology denotes revelation, the uncovering of that which was hidden. Over the last few decades, the sciences have revealed the extent of the hitherto unknown impact of human behavior and technologies on local and global ecologies, and the alarming future scenarios that might lie in wait. Faced with interlinked environmental crises such as climate change, ocean acidification, and rising extinction rates, we should heed Prince's words that we "can't run from revelation."[4] But there is a difference between responding to the frightening possible futures revealed to us by climatologists and believing that the end is nigh. The petrocapitalist party (to which many were never invited) might be coming to an end, but we are not, in Prince's words, "out of time." We need to counterbalance the doom of apocalypses, even ones with synth-funk backing tracks.[5]

Visions of likely or assured future destruction often sponsor cynicism and despair; they can therefore become self-fulfilling prophesies because they encourage a belief that all ameliorative action is futile.[6] Rather than orientating ourselves toward future disaster—or putting our faith in salvation through divine intervention, or through the invisible hand of the markets, or through an as-yet-undiscovered techno-fix—I suggest we shift from the apocalypse to come to the apocalypso now.[7] I first came across the word apocalypso in Evelyn Reilly's 2012 poetry collection *Apocalypso*, a book that explores environmental crisis in relation to cultural traditions of apocalypse.[8] Reilly never explicitly defines her suggestive title, but she does indicate the process that led her to it, explaining that her working title was *Apocalypse* but she began wondering "where's the joy?" and started "researching the emotions and practices of calypso music and changed the title."[9] I have adopted Reilly's portmanteau because it articulates the seriousness of the troubles that societies face without tipping into the destructive negativity that can arise from the purely apocalyptic. Apocalypso fuses the alarm and concern surrounding discussions of environmental crisis with the sense of play, togetherness, and critique typical of the calypso tradition.

The word *apocalypse* can refer to both a cataclysmic future event and the vision or text describing (i.e., revealing) that future—most famously the Book of Revelation, also known as the Apocalypse of John. My sense of apocalypso emerges from that second definition of apocalypse. The word apocalypso absorbs and modifies the word *apocalypse*. Accordingly, apocalypsos are texts or visions that absorb but also disrupt apocalyptic futures. Apocalypsos—whether they are poems, performances, or visions—may incorporate apocalyptic tropes and fears, but they are ultimately apocalypsoic (the adjectival form of apocalypso) in their resistance to apocalyptic pessimism. While my focus here is on apocalypso as vision or text, the dual meanings of apocalypse as both text/vision and event suggest a secondary definition of apocalypso: an event that resists an apocalyptic future via a fusion of joy and critique. This essay culminates in one such fusion: the defiant dancing at the 2011 Occupy Oakland protests.

Both senses of apocalypso resist apocalypse's inertial drag of cynicism and paralyzing hopelessness in the face of catastrophic futures. Where apocalyptic texts prophesize a disaster to come, apocalypsos reveal and revel in the possibilities of a troubled present. The state of the biosphere may mean that the present contains the seeds for a potential future apocalypse, but those seeds need not germinate.

Crisis may precede catastrophe, but it does not guarantee it. Cynical resignation that all is lost is a failure of responsibility and imagination. An apocalypso does not mean a calypso for a coming apocalypse; an apocalypso is not dancing at the end of the world because, despite its manifold troubles, apocalypsos do not see the world as about to end.[10]

Calypso is not just dance music. It evolved partly as extemporized, satirical Creole songs that were a form of resistance to and defiance of colonial authority in Trinidad. From the late nineteenth century onward, calypso songs grew much more elaborate, with a strong emphasis on lyrics as well as music and with performers (known as calypsonians) often taking aim at local social and political issues. Caribbean musicologists Peter Manuel, Kenneth Bilby, and Michael Largey explain that "penning verses about current events makes calypso a uniquely dynamic form of grass-roots folklore, closely attuned to people's daily lives."[11] As calypso developed, some calypsonians expanded their focus beyond the local, turning their critical eye to topics ranging from the treatment of Caribbean migrants upon their arrival in the United Kingdom to the technologies of American neoimperialism.[12] Although infused with fun, calypso is a politicized and critically aware tradition.

I do not wish for my description of apocalypso's fusion of apocalypse and calypso to co-opt or appropriate calypso's Caribbean specificity. Rather, I am interested in how a calypsoic combination of celebration, creativity, and critique might disrupt or negate the despair and reactionary cynicism associated with the ways we think and write about apocalypse. From the dancing protesters in Juliana Spahr's poetry collection *That Winter the Wolf Came* (2015) to the rave anthems that structure David Finnigan's satirical theatre project *Kill Climate Deniers* (2014–18), there are already multiple contemporary artworks that might well be described as apocalypsos, or that at least contain an apocalypsoic dimension.[13] Apocalypsos often share several features with calypsos. They demonstrate the capacity of the arts to synthesize and interrogate the sundry challenges of our current crises, and they also often maintain a link with tradition, recuperating what is valuable in the past in order to prepare for and cultivate a future that will not allow us to merely wallow in the negative. They celebrate fun—whether having fun or making fun. Apocalypsos know the importance of joy, but they are not naive; dancing in the face of disaster does not mean fiddling while the world burns. Instead, the word apocalypso speaks to an active and concerted effort to address our present troubles in all their complexity and creative possibility. As Naomi Klein suggests, our environmental crisis is a kind of "civilizational wake-up call" that functions as both a demand and an opportunity to tackle social and political issues: "we need an entirely new economic model and a new way of sharing this planet."[14] In the words of Margaret Atwood, "it's not climate change it's everything change."[15]

Reilly's *Apocalypso* illustrates how a text can bring together calypsoic celebration and apocalyptic anxiety. She explains that "we still need to embed ourselves in the joy of art, even when that art addresses potential disaster. So I hoped even for this grief-stricken book to stay tethered to that notion of music, of joy."[16] Part of the joy of the long title poem, "Apocalypso: A Comedy," is the way it plays with and transforms the apocalyptic tradition:

> This morning kicking against the pricks
> of wholesale legislative
> abandonment
>
> and in the distance the sirens
> add some lurid backup
> to these cataclysmic lyrics:
>
> *And a third of the sea became blood*
> *a third of the living creatures died*
>
> *And many were cast alive into a lake of fire*
> *and all the fowls were filled with their flesh*

(Note the touch of the vulture:
which, along with the condor,
is of the family Cathartidae
meaning *purifier*)[17]

By "cataclysmic lyrics," Reilly means both the four italicized lines that
are adapted from Revelation (8:8–9; 20:20–21) and the phrase "kicking
against the pricks." The latter is an allusion to a song Reilly references
several times throughout the poem: Johnny Cash's Judgment Day bal-
lad, "The Man Comes Around," where the Man in Black rasps, "It's
hard for thee to kick against the pricks."[18] To kick against the pricks,
a phrase with biblical origins (Acts 9:5), means to resist authority (as
an ox might "kick" against the "prick" of an ox-driver's stick). Reilly
therefore intimates resistance and outrage against a quiescent legisla-
ture that abandons its responsibilities, refusing to legislate and thereby
damning organisms and ecosystems, leading to the images of environmental despoliation that permeate the poem.

The italicized lines from Revelation and the accompanying nod to
avian purifiers highlight one of the central problems with the apocalyptic mode and the need for a countervailing apocalypsoic force. As
historian Matthew Avery Sutton and literary critic Marina Warner have
pointed out, apocalypses often encourage absolutist thinking that splits
the world into the damned and the saved.[19] Reilly focuses on this divisive aspect of apocalyptic tradition elsewhere in the poem:

disclosing things to certain entitled persons
things withheld from the majority
of humankind (Wikipedia
apocalypse definition two)

and allowing those with the seal
upon their foreheads
to torture the rest[20]

While the scorched landscapes of catastrophic climate change have usurped the "lake of fire" in the contemporary apocalyptic imaginary, the division between an enlightened elect and a doomed majority persists. We see this misanthropic ecological pessimism in figures such as the novelist and "recovering" environmentalist Paul Kingsnorth or the doomsday preppers made popular by the National Geographic TV series of the same name.[21] The danger is that the apocalyptic fears that drive individuals to prepare themselves and their immediate circle for an anticipated catastrophe mutates into a loss of hope for humanity in general. Conversely, apocalypsos promote solidarity and togetherness instead of individualist survival fantasies. This does not mean glibly asserting that "we are all in this together." From those island nations most threatened by sea-level rise to the disproportionate per-capita CO_2 emissions of the Anglosphere, neither the risks of nor the responsibility for environmental crises are evenly distributed. Nevertheless, apocalypsos emphasize the importance of (nonexclusive) community and cosmopolitan responsibility over a selfish insistence that since the end seems imminent, it's time to start building a bunker in the basement and stocking up on canned food and ammo.[22]

Apocalypsos resist the tragic inevitability that haunts many versions of apocalypse. It is worth dwelling on the etymological link that Reilly highlights between the flesh-consuming avian "purifier[s]" (*Cathartidae*) and tragic catharsis (i.e., the purgation of emotions via vicarious experience associated with Aristotle's description of tragic theater). This connection accrues significance as a result of Reilly's title: "Apocalypso: A Comedy." An apocalyptic tragedy might generate catharsis from readers experiencing trauma at a remove—the sort of vicarious revelry in another's difficulties we find in survivor narratives such as Cormac McCarthy's *The Road* (2006) or prepper-favorite William R. Forstchen's *One Second After* (2009). But where tragedies traditionally end with death, comedies frequently close with the coupling of characters and therefore the promise of future life. Apocalypsos are fundamentally comic (albeit a black comedy) in their affirmation of life's endurance. Fittingly, "Apocalypso: A Comedy" ends with a show of resistance to the apocalyptic tradition and an acknowledgment of a comic one:

> (You, too, John,
> should get some rest)
>
> Thank you friends, for your love and endurance
> This is the end of our revelatory revels[23]

By dismissing John of Patmos, Reilly tells us to give the overwhelming doom of apocalypse a rest. The "end of our revelatory revels" is a clear nod to Prospero's "our revels now are ended" speech from *The Tempest*, a hymn to the transformative power of the imagination ("we are such stuff / as dreams are made on") from a play where the apocalyptic storm is not the end but the beginning.[24] As she does throughout her text, Reilly emphasises the way the arts' disruptions and distortions can metamorphose literary and environmental apocalypses, turning them into something rich and strange that does not ignore crisis but that also refuses to succumb to despair—an apocalypso.

Juliana Spahr's poetic intervention into environmental crises provides an alternative example of what might constitute an apocalypso. The poems of Spahr's 2015 collection *That Winter the Wolf Came* move across interlinked political, economic, and ecological concerns, ranging from Occupy Oakland to Deepwater Horizon. Features such as the ominous tick-tock created by Spahr's continual insertions of Brent Crude's fluctuating spot price, or the images of exoskeletoned police and subterranean explosions could easily have been pressed into the service of a predominantly apocalyptic vision, but Spahr's focus is on participation, activism, and possibility in the face of seemingly overwhelming troubles. At some points the poems' dialogue between hope and despair erupts into an apocalypso. The following is from a poem fittingly called "It's All Good, It's All Fucked":

> walk out of the bar and down the street to the plaza to be with.
> And when I got to with, it was entirely possible, likely even,
> that Smooth Criminal was playing and a form of dancing
> that made no sense was going on, messy, chaotic, slightly
> frightening in its uneven physicality and very likely at that
> moment the sky was a deep, dark clear, with no stars because
> of the lights on the buildings. There jostled in that crowd
> by the felonious and the thieving and the sincere and the
> oppositionally defiant and the stoned and the overeducated and
> underemployed and the constantly shaking and the drunk all
> the time and the missing teeth and the bloodstained crescendo
> Annie and even by the socialist with the small yapping dog, at
> that moment I would feel I had made a right decision. Were
> we okay? Like Annie, of course we were not . . .[25]

This is apocalypso as event and as text describing that event—the revelation of the possibilities for resistance in the present. If the sky that Spahr's motley crew dances under seems sinister, it is perhaps because the disappearance of stars is itself an apocalyptic trope (e.g., Revelation 6:13 and Joel 2:10). The twice-used nounless preposition "with" suggests togetherness without specificity and therefore without exclusion. It is not the divisive apocalypticism of the damned and the elect but rather a state of openness to friend and stranger alike, not purification by fowl or fire but messy togetherness. The tone is likewise messy—which is to be expected when calypso meets apocalypse, fear meets fun, celebration meets critique. This is not a dance in avoidance of troubles faced but in acknowledgment of them. Facing rough and smooth (and sometimes perma-tanned, settle-out-of-court) criminals, like Annie, we are not okay. This can't go on; we must go on. A shared sense of risk engenders uncertainty, urgency, and experiment. Spahr swaps the despairing paralysis of a terrifying future for an active engagement with contemporary troubles.

Spahr's poem, like Reilly's, captures the spirit of apocalypso: fears for the future are transformed by imagination and fortitude into different ways of dreaming, making, and belonging in the present, thereby opening up different possible futures that are not such stuff as nightmares are made of. Cancel the apocalypse; apocalypso now!

NOTES

1. Prince, "1999," *1999* (Warner Bros., 1982).
2. Frederick Buell, "A Short History of Environmental Apocalypse," in *Future Ethics: Climate Change and Apocalyptic Imagination*, ed. Stefan Skrimshire (New York: Continuum, 2010), 14.
3. Prince, "1999."
4. Prince, "1999."
5. For an interesting discussion of Prince in relation to apocalypse and Will Smith's alternate millennial vision, see Malcolm Bull, "Tick-Tock," *London Review of Books*, December 9, 1999, https://www.lrb.co.uk/.
6. There is a counterargument that the shock of apocalyptic visions can function in a positive manner. For example, Mark Levene sees apocalypse "not as a prospect simply of obliteration, and with it world-end, but rather as a prophetic warning whose wake-up call to all humanity beckons them to participate in a general act of redeeming planetary reconciliation." See Mark Levene, "The Apocalyptic as Contemporary Dialectic: From Thanatos (Violence) to Eros (Transformation)," in Skrimshire, *Future Ethics*, 61.
7. A telling example of the curious belief that climate change will be solved by divine intervention is Republican congressman Tim Walberg. See Mahita Gajanan, "Republican Congressman Says God Will 'Take Care Of' Climate Change," *Time*, May 31, 2017, http://time.com/.
8. There is an earlier—and, outside of poetry circles at least, much more famous—

coining of the word apocalypso as the title of electro duo The Presets's platinum album *Apocalypso* (Universal/Island, 2008).

9. Katja Jylkka, "Discussing the Value of Ecopoetics with Evelyn Reilly," UC Davis Humanities Institute, http://dhi.ucdavis.edu/.

10. Interestingly, though, a calypso for the apocalypse has been written. See Graham Roos, *Apocalypse Calypso* (Buckingham: University of Buckingham Press, 2012).

11. Peter Manuel, Kenneth Bilby, and Michael Largey, *Caribbean Currents: Caribbean Music from Rumba to Reggae* (Philadelphia, Pa.: Temple University Press, 1995), 195.

12. See, e.g., the calypso "Satellite Robber"—"I'm here to rip out your heart, tear your culture apart, make you worship the American flag. . . . don't aggravate me, when it coming to TV, it is I who control you" (quoted in Manuel, Bilby, and Largey, *Caribbean Currents*, 195) and the analysis of "Signal Da Plane" in Kezia Page, "'Everybody Do the Dance': The Politics of Uniformity in Dancehall and Calypso," *Anthurium* 3, no. 2 (2005).

13. Juliana Spahr, *That Winter the Wolf Came* (Oakland, Calif.: AK Press, 2015). For a description of Finnigan's project, see Kill Climate Deniers, http://www.killclimatedeniers.com/. Another strong theatrical contender for apocalypso status is Bruno Latour's play *Gaia Global Circus*, which toured in English and French from 2013 to 2015. For details, see Bruno Latour's website (http://www.bruno-latour.fr/).

14. Naomi Klein, *This Changes Everything: Capitalism versus the Climate* (New York: Simon & Schuster, 2014), 25.

15. Margaret E. Atwood, "It's Not Climate Change—It's Everything Change," *Medium*, July 27, 2015, https://medium.com/.

16. Evelyn Reilly, "Evelyn Reilly in Conversation with Andy Fitch," Something on Paper, http://www.somethingonpaper.org/; originally published in *60 Morning Talk* (New York: Ugly Duckling Press, 2014).

17. Evelyn Reilly, *Apocalypso* (New York: Roof Books, 2012), 85. Italics in original.

18. Johnny Cash, "The Man Comes Around," *Johnny Cash IV: The Man Comes Around* (Universal, 2002).

19. Matthew Avery Sutton, *American Apocalypse: A History of Modern Evangelicalism* (Cambridge, Mass.: Harvard University Press, 2014); Marina Warner, "Angels and Engines: The Culture of Apocalypse," *Raritan* 25, no. 2 (2005): 12–41.

20. Reilly, *Apocalypso*, 101. Italics in original.

21. *Doomsday Preppers* (2011–14), https://www.nationalgeographic.com.au/tv/doomsday-preppers/; Paul Kingsnorth, *Confessions of a Recovering Environmentalist* (London: Faber & Faber, 2017). See also the useful discussion in chap. 1 of Matthew Schneider-Mayerson, *Peak Oil: Apocalyptic Environmentalism and Libertarian Political Culture* (Chicago: University of Chicago Press, 2015).

22. For an introductory insight into preppers and the items they are likely to hoard, see Rod O'Connor, "These Suburban Preppers Are Ready for Anything," *Chicago Magazine*, May 2014, https://www.chicagomag.com/.

23. Reilly, *Apocalypso*, 110.

24. William Shakespeare, *The Tempest* (1611), new ed., ed. Cedric Watts and Keith Carabine (Hertfordshire: Wordsworth, 1994), 4.1.138–49.

25. Spahr, *That Winter the Wolf Came*, 71.

Blockadia

Randall Amster

Pronunciation: bla-kaid-ee-a (blɒ:keɪdɪə)

Part of Speech: Noun

Provenance: Activism

Example: The machine made its way across the stark landscape, grinding a series of earthen scars into decimated habitats. Yet the people who had assembled adjacent to the site were more than mere witnesses to the devastation; they were there to bring it to a halt. They were no more than a score, standing in stark contrast to the sheer tonnage of heavy equipment and the intersection of private security and law enforcement on site to keep the peace. One by one, the members of the assembly linked arms and snaked their way across the path of the leviathan. It belched smoke and churned dust as it proceeded toward them, but the human chain obstructed the road and forced the machine operator to halt his progress. The authorities attempted to remove the activists, who were linked together with a bricolage of ropes, wires, and concrete that would take hours to disentangle. Meanwhile, another protest camp was forming a few hundred feet down the road, preparing to do the same. Welcome to Blockadia.

Modern society is marked by a thirst for fossil fuels. To continue to obtain these resources, it becomes necessary to drill deeper, pump harder, and transport fuels across greater distances over precarious terrain. Communities and ecosystems alike are inevitably affected by these activities, and sometimes these impacts precipitate a tipping point sufficient to spark resistance. A prominent example is the Keystone XL pipeline, which was planned to connect the bitumen tar sands in Alberta, Canada, to a junction point in Steele City, Nebraska, and then to refineries in Texas and elsewhere. Its construction has been dogged by controversy, and in 2011 and 2012 a mounting activist campaign to disrupt its completion expanded to engage larger environmental and climate issues as well as Native American objections over sovereignty and land rights. Numerous protests and demonstrations at the national level were launched by indigenous and climate justice activists in Washington, D.C., in an attempt to convince lawmakers and the Obama administration to reject plans to complete the pipeline. These actions included rallies, marches, civil disobedience, and other forms of symbolic nonviolence focusing on government officials. As a result of this pressure, President Obama rejected approval of the Keystone XL pipeline in 2015, only to have his decision reversed by President Trump in 2017.

The struggle over the Keystone pipeline represented a watershed moment in the trajectory of modern North American environmental activism. Concurrently, the grassroots movement Idle No More was founded by three First Nations women in Canada in December 2012, prompting a series of actions drawing together issues of sovereignty, environmental preservation, and social justice. In 2013 Melanie Jae Martin and Jesse Fruhwirth highlighted the emergence of local resistance to pipelines and related projects, including the use of nonviolent direct action as a tool. Their article, "Welcome to Blockadia" (constituting the first use of the term in print), valorized Blockadia as "a place where the future of the environmental movement is being negotiated." Three particular actions were cited as examples of the concept: a march to a site along the Keystone pipeline route to kick off a campaign against its construction; a demonstration that involved "storming" the corporate offices of the pipeline's primary architect; and a "tree-sit" along the pipeline's planned construction path. These actions were decentralized yet focused on the same extraction project, representing a convergence of tactics undertaken by a diverse network of participants. As Martin and Fruhwirth observe, this confluence captured the essence of the term: "These actions represent the spirit of Blockadia—a vast but interwoven web of campaigns standing up against the fossil fuel industry and demanding an end to the development of tar sands pipelines."[1]

Martin and Fruhwirth cited numerous factors that established Blockadia as a new paradigm in mainstream North American environmentalism, including "the normalization of direct action; the involvement of rural and indigenous groups along with more typical 'activists'; and the ability of tar sands extraction to motivate even those who tolerated conventional oil pipelines." They concluded that the rise of Blockadia was "building a unified front" within the environmental movement—and they contended that by doing so, it was "making the struggle potentially winnable despite the steep odds against it." The rationale was that Blockadia not only offered an opportunity to block specific projects but also could radicalize environmentalists, "with historically marginalized people stepping into the forefront of the movement, and historically privileged groups fighting for rights they once took for granted." Given the importance of these issues and the diversification of the actors involved, these convergent factors helped prompt greater awareness of "the intertwined nature of social justice and environmental issues" and, critically, was inspired by the "active engagement of people in the frontline communities."[2]

After the term's introduction, activist and author Naomi Klein began applying the concept more widely, likewise attributing its coining to "the movement against the Keystone XL Pipeline in Texas [by] the people who are blocking the fossil fuel projects with their bodies and in the courts and in the streets."[3] In a 2014 *Democracy Now!* interview, Klein characterized Blockadia as "transnational space, roving space";

in her popular book *This Changes Everything* (2014), she reinforced the notion that Blockadia "is not a specific location on a map but rather a roving transnational conflict zone that is cropping up with increasing frequency and intensity" to confront extractive projects.[4] What unifies these actions is the industrial impetus to continue mounting deeper environmental incursions through more unconventional and riskier means of pursuing profit. Blockadia might be viewed as a parallel set of interventions, going deeper as the stakes and scale of the challenges escalate, and doing so through riskier and less conventional methods.

The range of shared tactics being utilized—including "packing local council meetings, marching in capital cities, being hauled off in police vans, even putting their bodies between the earth-movers and earth"—is mirrored by the diversity of the actors involved, who generally "look like the places where they live, and they look like everyone: the local shop owners, the university professors, the high school students, the grandmothers."[5] Inspired by indigenous activists around the globe, contextualized within the unique geographies of local communities, and informed by the successful historical examples of nonviolent social change, Blockadia draws our gaze to the critical struggles of our time, which threaten to impinge on our "collective survival" as beings dependent on access to healthy soil, clean air, and fresh water.[6] Blockadia isn't merely about better access to resources or securing narrow rights. It also calls into question the baseline operations of the dominant political and economic paradigm. In this sense, Blockadia is the space between the world as we find it and the world as it should be.

Blockadia thus can be viewed as both a response to environmental damage and a collective survival strategy manifested in particular places by diverse peoples. Blockadia emerges at the same moment that ecological thresholds of biodiversity and climate stability are being approached, regulatory firewalls are in danger of being breached, and life-giving capacities are being damaged on a global scale.[7] As such, Blockadia merits a place in the lexicon of the Anthropocene, with its direct applicability to movements focusing on climate change, environmental degradation, and the impacts of these forces on ecosystems and human communities alike. This isn't simply a political or economic movement, or even primarily a cultural or ethical contest. Rather, it makes an existential intervention that demands a future, reflective of the mantra "Respect existence, or expect resistance." It might be said that Blockadia is what the world looks like when ordinary people are called to extraordinary measures. When the mechanisms of governance and oversight have failed them, people can be compelled to utilize the potent mechanisms of collective action and civil resistance, sometimes with their bodies as the vehicle for their message.

More than simply comprising a set of tools or techniques for organizing and promoting change, Blockadia indicates a state of mind that fosters both a critique of stratified distributions of power and a reinvigoration of the responsibility and ability to confront them. The invocation of the term is consistent with emerging methods of movement organizing that are ostensibly decentralized yet also unified in their emphasis on redressing the structural factors that undergird contemporary crises, serving to highlight the need for urgent action as an antidote to complicity and complacency. Bill McKibben, founder of 350.org and a leading climate advocate, goes so far as to view this struggle through the lens of a war that must be won if humankind is to survive.[8] In waging it, we may discover a capacity to resist and struggle as well as a renewed spirit of engagement and commitment to being the architects of a better future.

Yet even as we look ahead to a world animated more by creativity than calamity, we are reminded of the lessons of the past. Civil disobedience has been deployed as an instrument of environmental change for decades, but the range of actors and the scope of action are important additions. As Sarah Tory notes, this movement isn't "just about fighting the next pipeline project or coal terminal, nor is it driven solely by West Coasters' climate change concerns." Instead, Blockadia is the leading edge of a "swelling tide of calls for indigenous rights," bringing forth "a renewed emphasis on social justice to the environmental movement."[9] The demonstrations and encampments against the Dakota access pipeline at Standing Rock, North Dakota, manifested this convergence of lessons from the past, struggles in the present, and visons of and for the future. Standing Rock galvanized contemporary organizing by advancing the merger of sociopolitical and environmental concerns. It also marks a potential passageway through the liminal region between the volatility of today and a more inspired tomorrow in which humankind finds its fullest expression as part of natural processes rather than as their antagonists. Blockadia isn't on the map; it is the map, showing us a path forward, together.

The linkages being forged in these moments are potent mechanisms for connecting our histories and struggles, and for "jumping scale" from local interventions to global networks of resistance.[10] Assessing the legacy of Standing Rock as the campaign continued despite political setbacks, Sarah van Gelder observes that "a revolution in values and culture is rippling out across the country and the world," animated by changes in consciousness and reformulations of power.[11] Along with the rising challenges engendered by the Trump administration and others with a similar penchant for climate denial and rights deprivations, a new era of resistance has emerged, grounded in a justice-oriented perspective that calls on environmentalism "to defend a living world that is under assault at every point, from the global climate to the most vulnerable communities."[12] In this formulation, ecology is a practice as well as a body of knowledge, and politics are necessarily proactive. "Economic power, racial inequality, and the struggles of indigenous peoples are not optional or supplemental," writes Jedediah Purdy. "They are at the heart of the work."[13] Blockadia is as mundane as how we navigate daily life, and as visionary as who gets to decide.

In his pointedly crafted collection *Once in Blockadia*, poet and activist Stephen Collis reminds us that "our selves," "the ones we have been waiting for," are the architects of "other possible futures." Like a beacon in the dark, "we are engines of change, component parts, aqueducts," reaching "back in time and forward in time, lifting materials from the forest to be a barrier to human stupidity."[14] In this lyrical vision, we seek "plausible futures" in a moment of profound uncertainty and rampant change, finding ourselves at a crossroads where "time and growth" seem to diverge (with the former running out while the latter seeks to expand). Here, "paths emerge" toward a future that is perhaps "cleaner" in its production and consumption yet will also be turbulent from the impacts already accrued—ultimately yielding a storyline in which "environment play" (in which attitudes and actions are simultaneously reoriented) helps "shape the turbulence" in such a way that "tomorrow's world might navigate the inevitable."[15] In this sense, Blockadia doesn't posit that we can or will avert cataclysm but rather that we rededicate ourselves to surviving it.

Blockadia points toward a world in which people find themselves bombarded by seemingly insurmountable environmental and sociopolitical crises yet nonetheless choose to forge ahead with a spirit of determination and possibility. As a conceptual touch point for bringing these forces into sharper focus, Blockadia offers a nascent synergy of attitudes and actions, of values and visions, of righteous rage and eternal optimism—core elements for reconnecting humankind to the overarching web of life, and for sustaining the life-giving properties of ecosystems that make human societies possible at all. The essence of Blockadia is more than simply imagining a better future; it entails remembering the lessons of history and developing tools to resist sociopolitical injustice and ecological degradation in the present.

Ildsjel ANOTHER PATH

NOTES

1. Melanie Jae Martin and Jesse Fruhwirth, "Welcome to Blockadia!," *Yes! Magazine*, January 11, 2013, https://www.yesmagazine.org/.
2. Martin and Fruhwirth, "Welcome to Blockadia!"
3. Joshua Holland, "Naomi Klein: Only a Reverse Shock Doctrine Can Save Our Climate," Moyers, September 16, 2014, https://billmoyers.com/.
4. "Naomi Klein on the People's Climate March and the Global Grassroots Movement Fighting Fossil Fuels," Democracy Now!, September 18, 2014, https://www.democracynow.org/; Naomi Klein, *This Changes Everything: Capitalism versus the Climate* (New York: Simon & Schuster, 2014), 254.
5. Klein, *This Changes Everything*, 254.
6. Klein, *This Changes Everything*, 255.
7. For example, see Will Steffen, Katherine Richardson, Johan Rockström, et al., "Planetary Boundaries: Guiding Human Development on a Changing Planet," *Science* 347, no. 6223 (2015): 1259855.
8. See Bill McKibben, "A World at War," *New Republic*, August 15, 2016, https://newrepublic.com/.
9. Sarah Tory, "Dispatch from Blockadia," High Country News, January 20, 2016, https://www.hcn.org/.
10. Compare Patrick Bond, "Challenges for the Climate Justice Movement: Connecting Dots, Linking Blockadia, and Jumping Scale," EJOLT Report 23: Refocusing Resistance for Climate Justice, 2015, http://www.ejolt.org/.
11. Sarah van Gelder, "How Standing Rock Has Changed Us," *Yes! Magazine*, December 7, 2016, https://www.yesmagazine.org/.
12. Jedediah Purdy, "Environmentalism Was Once a Social-Justice Movement," *Atlantic*, December 7, 2016, https://www.theatlantic.com/.
13. Purdy, "Environmentalism Was Once a Social-Justice Movement."
14. Stephen Collis, *Once in Blockadia* (Vancouver: Talonbooks, 2016), 17–18.
15. Collis, *Once in Blockadia*, 41–55.

Cibopathic

Daniel Worden

Pronunciation: see-boh-path-ik (siːbəʊːpaθɪk)

Part of Speech: Adjective, noun (cibopath)

Provenance: Speculative comic book series (John Layman and Rob Guillory's *Chew*)

Example: An individual with cibopathic abilities "can take a bite out of an apple and get a feeling in [her] head about what tree it grew from, what pesticides were used on the crop, and when it was harvested."—John Layman, 2017

The labels in upscale supermarkets today—"natural," "organic," "grass fed," "free range," "GMO free," and "locally sourced"—are meant to tell us how a piece of food was grown, harvested, processed, or transported. Yet as anyone concerned with food production and food culture today knows, these categories are often misleading in their connotations, implying a local origin to corporate foodstuffs, or making the products of small local farms seem less sustainable because those farms cannot raise the capital to pay for expensive organic certification. That these slogans are common both to greenwashed industrial food products and to local farmer's markets makes it clear that contemporary food movements are related to, yet not always synonymous with, contemporary environmentalism. The ability of the cibopath offers a way to speculate about how these food categories, and the struggles they refer to might be figured as both personal and political. What would it mean if we had no choice but to taste history? Would sustainable food taste differently than unsustainable food? Can gastronomic pleasure serve as a conduit for environmentalist thinking, feeling, and action? Cibopathic abilities are both fictional projections that imagine answers to these questions and exaggerations of real-world phenomena, like supertasting and the nose-to-tail cooking philosophy, that are reorienting our relation to food.

Food has long been a site of struggle. That struggle has taken on new valences in an era of industrial farming, processed foods, and global logistics that deliver produce from around the world to our supermarkets, displacing seasonal patterns of eating. From fine-dining pioneers like Alice Waters and Dan Barber to Walmart and Target's increased attention to organic foods in a bid to compete with the Whole Foods supermarket chain, the documentation and labeling of where food comes from and how it is produced has become a way to imagine more sustainable and also more delicious food, as well as a way to market expensive alternatives to processed foods as better for the consumer's health and the environment. Stemming from both nutritional and culinary interests, the push toward more organic and local foods reflects a larger cultural shift, one that Wendell Berry recorded in his influential 1977 book *The Unsettling of America*, in which Berry critiqued industrialized agriculture and the consumer culture it grew out of. As Americans became more suburban, and therefore more removed from sites of food production, Berry argues that the ties between agriculture and culture were remade. The result of this transformation is the full subsumption of farming into market logic—or, as in Berry's account, an increasing distance between the farmer and the consumer, as agriculture becomes more industrialized and the population becomes more urban and suburban. As a result, "the consumer eats worse, and the farmer farms worse."[1] This market logic and its concomitant industrialization of agriculture has reduced the kinds of food shortages and price spikes that previously determined food struggles such as famines and riots in the Global North. Yet as many studies have pointed out, the industrialization of agriculture and the full commodification of food-stuffs has led to a steady depletion of soil quality and nutritional content, along with natural flavors, as what we might think of as farming practices have been replaced with industrialized and chemical processes.[2]

Today one of the major problems that those of us privileged with access to food face is how to make food available to everyone, something that industrial agriculture was designed to do with its massive crop yields and monocultures. Ensuring that food is produced in ways that sustain the soil as well as the bodies of those who consume it seems to be something that industrial agriculture is unsuited to accomplish for the very reasons that have led to its dominance. As legal scholar and environmentalist thinker Jedediah Purdy has noted, there is a diverse "food movement" today that "has no organized center. It shows up as an interest in where food comes from, who grows food and how, and the way food travels from farm to plate. It is evident in consumer fads and high-end restaurants, local economies that have been rebuilt around community-supported agriculture and farmers' markets, and people's renewed eagerness to put their own hands in the dirt. Altogether, it hints at a new picture of people and nature."[3] Key texts in this contemporary food movement include Michael Pollan's *The Omnivore's Dilemma* (2006) and Eric Schlosser's *Fast Food Nation* (2001), documentary films like *Food Inc.* (2008) and *Super Size Me* (2004), and a number of cookbooks that encourage simple and seasonal cooking, such as Deborah Madison's *Local Flavors* (2002), Ian Knauer's *The Farm* (2012), and Alice Waters's *The Art of Simple Food* (2007). The promise of the food movement is that it makes visceral the sometimes abstract ecological ideas attached to the discourses of sustainability, climate change, and environmentalism, such as shifting consumer habits from global food distributors to local farms, returning to seasonal eating patterns rather than expecting all kinds of produce to be available year round, and reducing food waste by consuming parts of plants and animals that are typically discarded, like carrot tops or chicken hearts.

When it comes to food, one hopes, you can taste the difference between sustenance and depletion, so that eating more sustainably coincides with eating more deliciously.

Advocates for local and organic food cultures often make a dual argument: local, sustainable agriculture is better for both public health and the environment; and food produced locally and sustainably simply tastes better. Sustainable and local farming practices are far more varied than the massive monoculture methods common in large industrial fields. The taste of fresh, local vegetables and free-range meat is more varied and subtle than the now familiar salty, fatty, and sugary flavors manufactured in industrially processed foods. Yet producing more local, more sustainable, and better food has become difficult in the United States because of the power of what food advocate Michael Pollan has described as Big Food—the cluster of food corporations that control vast amounts of food production in the United States, from seed stocks to restaurant supplies. As Pollan notes, Big Food is shockingly consolidated, controlling well over 50 percent of the American food supply and up to 80 percent of foods like beef: "Simply put, [Big Food] is the $1.5 trillion industry that grows, rears, slaughters, processes, imports, packages and retails most of the food Americans eat."[4] As Pollan and other food advocates have documented over the past twenty years, Big Food engages in practices that lead to environmental harm through soil depletion, unnecessary use of pesticides, genetic modification of crops, and privatization of seed stocks. It also encourages food waste through profligate restaurant and supermarket practices, which result in health problems such as diabetes and obesity, caused by the use of corn syrup and other additives in readily available processed foods. These practices and their harmful effects are widely acknowledged, yet they have not resulted in public policy changes because of the lobbying power of the food and agriculture industries. The pressing question today remains how we can turn our food system upside down, so that our consumption of food does not contribute to the steady decline of life on the planet.

The concept of the cibopath is a useful way of framing an individu-alized yet also scalable solution to this impasse. What if we could taste the history of our food? What if its production process was sensible to us? What if biting into something produced a visceral epiphany about the way our food was grown, how it was harvested and processed, and how it was cooked? If we tasted history instead of bitter, sweet, salty, sour, and umami, what would be delicious? What food wouldn't taste like oil, owing to the petroleum by-products used in fertilizer and pes-ticide, the fuel burned transporting it from farm to factory to table, and the fuel expended in refrigeration and cooking? Could an industrially farmed apple ever just taste like "apple"?

The term cibopath comes from the comics series *Chew*, written by John Layman and drawn by Rob Guillory.[5] *Chew* takes place in a near-future world where a bird flu pandemic has killed millions. Chickens and other poultry products have been outlawed, and the U.S. Food and Drug Administration has become a major policing force. The series' main character, Tony Chu, is a cibopath; when he eats anything, he experiences the history of that thing. As an FDA agent, Tony Chu solves food-related crimes by eating. While *Chew*'s often humorous content (starting with the literalness of its main character's ability to "take a bite out of crime") has been expanded into a complicated serial world since the comic series began in 2009 (it concluded with issue 60 in November 2016), *Chew* also explores some of the ways in which the food supply in North America is policed, governed through institu-tions, and manufactured on a level that has reshaped agriculture. Com-ics might seem to be a strange place to find a framework for thinking about the politics and possibilities of sustainability. Yet in recent years comics have sought to represent the food movement and its effects in mainstream media in sophisticated ways.[6] *Chew* is not alone, then, as a comics commentary on the food movement, though its central conceit of cibopathy makes it a more experimental take on the possibilities of the food movement.

In the first issue of *Chew*, detective Tony Chu's powers are made visible in a way that is uniquely suited to the comics medium. When he takes a bite of something, a vast grid of objects, actions, and materials cascades behind him on the page. Guillory's representation of Chu's cibopathy emphasizes its immediacy, as he experiences a cognitive surge of information.[7] As the story develops, Chu learns to refine his powers, so that by the end of the sixty-issue series, he not only is able to experience the history of whatever he eats but also has taken on other superpowers from those he has eaten (such as contortionism and the ability to tie any knot). Cibopathy makes an object's history into memory through ingestion, and that new memory changes the way Chu acts, thinks, and eats. History carries with it embodied practice, and our sensorial connections to food are saturated with habitual practices, cultural histories, and memories. In her study of matsutake mushrooms and their roles in global food networks, for example, anthropologist Anna Lowenhaupt Tsing finds in the smell of the matsutake a series of questions about history, food production, and economic networks. A key element of flavor, the mushroom's scent, launches her intellectual engagement with food, economics, labor, and culture: "At my first whiff, I was just . . . astonished. . . . My surprise was not just for the smell. What were Mien tribesmen, Japanese gourmet mushrooms, and I doing in a ruined Oregon industrial forest?"[8] This revelry leads her to view the matsutake and its unique history as both existing within and resonating with possibilities outside of capitalist modes of production. Similarly, *Chew*'s cibopath holds out the promise of being able to taste political economy, soil health, animal welfare, and pollution.

In *Chew*, food-related abilities proliferate. These include the Cibolocutor, who can communicate discrete messages and works of literature through food; the Saboscrivner, who can write about food so eloquently that readers experience tasting her words; and the Cibovoyant, who can see the future of anything she eats. All in all, *Chew* makes literal the ways in which food is as much a cultural force as a natural resource. This is made most emblematic in the series as the narrative shifts from a police procedural, with Chu joining the FDA and tracking down illegal chicken operations, to a science fiction story about an apocalypse brought on by the overconsumption of poultry. *Chew* makes a logical step from food history to the long-term viability of human life on the planet Earth.

Indeed, as many members of the sustainable food movement emphasize, changing our food culture is necessary in light of climate change, expected population growth, and the environmental effects of industrialized agriculture. Like the fossil fuels that underwrite its circulation, fertilization and preparation of food is artificially inexpensive to many in North America and elsewhere, and it has given rise to a consumption-heavy lifestyle that simply cannot be sustained for many more generations. As the U.S.-based nonprofit Environmental Working Group has argued, changing consumption and waste habits would help to distribute food more evenly across the globe and help to prepare agricultural systems for the population growth that is expected in decades to come.[9] Yet the U.S. food system today is markedly wasteful in its focus on meat production and consumption, as well as its reliance on monocultures. Perversely, our food system's reliance on fertilizers, pesticides, and genetically modified crops to resist the diseases that can devastate monocultures might be more effectively countered with more diverse crops and an expanded consumer palate for a range of vegetables and grains.

In *The Third Plate* (2014), chef Dan Barber argues that our consumption habits must change and that fine dining can and should provide a kind of avant-garde for newly responsible and sustainable food. An early icon of the farm-to-table movement in fine dining, Barber uses ingredients from his family farm and has worked with agricultural scientists to breed vegetables for flavor and not for yield. These experiments, in Barber's account, result in flavors that are unparalleled in typical supermarket produce. In Barber's book, descriptions of tasting real food, prepared with real ingredients, often spark recollections and comparisons across time and ingredients. For example, Barber describes tasting brioche made with fresh-ground wheat:

> The brioche *was* delicious, comforting in the way bread should be, but also a little exciting, with a flavor of toasted nuts and wet grass. Just as the Eight Row polenta tasted of corn—reminded me (because I needed to be reminded) that dried corn should actually taste of corn—the whole wheat brioche tasted distinctly of wheat. . . . The experience reminded me of my first taste of raw milk. . . . I couldn't believe what I was tasting. It was creamy and sweet, but also tangy, with a scent that reminded me of morning pasture. . . . It hardly resembled the version I knew before.[10]

Barber's descriptions of flavor—as being surprisingly rooted in the places where plants are grown and cattle graze—imagines the sensation of taste as having a cibopathic quality. *Chew* makes this refined palette into a superpower, yet for Barber it is, importantly, something that we all have access to, something that we can all refine and sharpen. Indeed, psychologist Linda Bartoshuk estimates that about 25 percent of the human population are supertasters; their tongues have more fungiform papillae, and they therefore have heightened responses to bitter tastes. Supertasters seem to have an aversion to fatty and salty foods; meanwhile, they also are averse to bitter-tasting vegetables (so it's not all roses). Bartoshuk's research and other work on early childhood taste sensitivity has demonstrated that a heightened sensitivity to taste and exposure to varied foods at any early age might explain some eating habits and taste preferences.[11] Infants exposed to more foods at an early age might be able "to trust new foods later in life" and thus be more equipped to eat seasonally rather than rely on favorite foods available year round only if they are shipped across the globe.[12]

A supertaster extraordinaire, the cibopath tastes the way Dan Barber wants us all to taste. And while in *Chew* cibopathy is often unpleasant—the history of a murder victim or a fast-food hamburger patty are equally traumatic—it raises the question of what an enjoyable experience of cibopathy would be like. What if a delicious plate of organic vegetables simply tasted like leaves, stalks, roots, sunlight, soil, and water? What if bread tasted distinctly of hull, bran, and endosperm? What if meat tasted like grass, acorns, and blood? For those able to afford to shop at farmer's markets, these tastes are more attainable. The political problem, though, is making these taste experiences and these histories available. As *Chew*'s bracing conclusion emphasizes, when human-size chickens arrive on Earth, ready to destroy the planet if poultry is still being eaten, our eating habits might prove intractable and ruinous. Everything tastes like exploitation. Could it be otherwise?

The concept of the cibopathic offers us the ability to think of food production as a sensory history. We perhaps too readily think of taste as a product of individual exertion (a master chef at work in the kitchen), global capitalism (a Big Mac tastes the same in Baltimore as it does in Bangkok), or mere nature (apples to apples). What if these three things tasted like the processes they are, the histories they drag behind them, and the networks required to sustain them? If we could all be cibopathic, we might be able to taste the difference between exploitation and sustenance. Cibopathy is one way of describing what the food movement has aspired to activate in all of us: the hope that in reflecting on our tastes and habits, we connect gastronomic pleasure to modes of producing food that replenish the Earth.

NOTES

1. Wendell Berry, *The Unsettling of America: Culture and Agriculture* (Berkeley, Calif.: Counterpoint, 2015), 42.
2. For recent accounts of soil depletion due to industrial agriculture and the consequences of shifting to processed foods that are flavored with additives, see Lindsey Haynes-Maslow and Jeffrey K. O'Hara, *Lessons from the Lunchroom: Childhood*

Godhuli
ANOTHER PATH

Obesity, School Lunch, and the Way to a Healthier Future (Berkeley, Calif.: Union of Concerned Scientists, 2015); and *Rotating Crops, Turning Profits: How Diversified Farming Systems Can Help Farmers while Protecting Soil and Preventing Pollution* (Berkeley, Calif.: Union of Concerned Scientists, 2017), https://www.ucsusa.org/.

3. Jedediah Purdy, *After Nature: A Politics for the Anthropocene* (Cambridge, Mass.: Harvard University Press, 2015), 232.

4. Michael Pollan, "Big Food Strikes Back," *New York Times Magazine*, October 9, 2016, 44.

5. John Layman and Rob Guillory's *Chew* was a sixty-issue comics series published by Image Comics from 2009 to 2016. It has been collected into twelve trade paperback volumes as well as larger hardback collections.

6. Series like *Get Jiro!*, a near-future story about a rebellious sushi chef written by celebrity food writer and TV personality Anthony Bourdain, and *Starve*, about an extreme reality TV cooking competition, both focus on celebrity chefs and their association with foodie culture, and they both take as their protagonists chefs who prize traditional methods and ingredients over fussy, fancy fine dining. More recent comics series that focus on chefs and sustainability in speculative worlds include *The Dregs, Flavor, Umami,* and *Chew* artist Rob Guillory's *Farmhand*. See Anthony Bourdain, Joel Rose, Langdon Foss, and José Villarubia, *Get Jiro!* (New York: DC/Vertigo, 2013); Rob Guillory, *Farmhand* (Berkeley, Calif.: Image, 2018); Joseph Keatinge, Wook Jin Clark, Tamra Bonvillain, Ariana Maher, and Ali Bouzari, *Flavor* (Berkeley, Calif.: Image, 2018); Ken Niimura, *Umami* (Panel Syndicate, 2018), http://panelsyndicate.com/comics/umami; Zac Thompson, Lonnie Nadler, Eric Zawadzki, and Dee Cunniffe, *The Dregs* (Los Angeles: Black Mask, 2017); Brian Wood, Danijel Zezelj, and Dave Stewart, *Starve*, 2 vols. (Berkeley, Calif.: Image, 2016).

7. For an account of how taste functions as cognition, see John S. Allen, *The Omnivorous Mind: Our Evolving Relationship with Food* (Cambridge, Mass.: Harvard University Press, 2012).

8. Anna Lowenhaupt Tsing, *The Mushroom at the End of the World: On the Possibility of Life in Capitalist Ruins* (Princeton, N.J.: Princeton University Press, 2015), 14.

9. Emily Cassady, "Feeding the World without GMOs," Environmental Working Group, March 2015, http://www.ewg.org/.

10. Dan Barber, *The Third Plate: Field Notes on the Future of Food* (New York: Penguin, 2015), 335–36.

11. For a summary of some of this research, see Mary Beckman, "A Matter of Taste," *Smithsonian Magazine*, August 2004, http://www.smithsonianmag.com/.

12. Beckman, "Matter of Taste."

Dàtóng

Andrew Pendakis

Pronunciation: da-tong (dɑtɔŋ)

Part of Speech: Noun

Provenance: Confucianism

Example: The age of dàtóng is at once long passed and still yet to come.

Dàtóng (大同) is a concept derived from the Confucian tradition that is usually translated in English as "great universality" (and sometimes "great harmony" or "great unity"). One of the earliest recorded usages of the term appears in *The Book of Rites*, a text on ceremonial practice that does not assume its present form until the Han Dynasty (206 BC to AD 220). Invocations of dàtóng echo across the long arc of Confucianism. They reappear in the work of the late Qing (1644–1912) reformers and can still be found in the thought of China's great twentieth-century theorists, including Li Dazhao and Sun Yatsen. Though the term is never used directly by Confucius himself, the spirit of the concept signaled by the word is present everywhere in his work.

The best way to understand the place of dàtóng in the Confucian tradition is along the lines of the role played by the concept of the *kallipolis* (ideal city) in Plato: both are signs for an ideal configuration of persons and things—words, as it were, for true, ideal, or correct "political" order. This last term uses quotation marks to foreground its inadequacy: both traditions implicitly reject the notion of a domain intrinsic to the political (where people do properly "political" things like vote and pass legislation), merging politics with moral, aesthetic, and even onto-logical considerations. Dàtóng is the state of affairs in which humans become (actually) what they are (potentially). Dàtóng is, in other words, the collective practice that matches humans to their own lost essence and in so doing harmonizes them with the essence of things them-selves. It may be that for Confucius this model was never more than a pragmatic ideal, a spur to moral effort rather than a fully imagined and achievable social model. Conceived of as such, dàtóng would func-tion within the space of the political in the same way as the concept of the sage (聖) does in the purview of individual moral practice. The sage or *sheng* is the perfectly just individual, a figure that Confucius openly claims to have "no hope of [ever] meeting."[1] However, this understand-ing of dàtóng as no more than a moral spur is complicated by the fact that Confucianism—unlike Platonism, for example—locates its ideal future in an actually existing past: the question of whether or not uto-pia might exist was hereby offset by the fact it already had. Dàtóng had actually existed, and not in a murky, preternatural past—some vague Chinese Atlantis—but at a precise moment in history: the reign of the great Zhou emperors of the eleventh century.

For Confucius, social order is the direct consequence of virtue or *ren* (仁). Where rulers, families, and individuals have abandoned the *dao* of right practice, disorder, poverty, and suffering prevail. Chaos, on these terms, represents the domination of immediate desire over the time-tested patience of moral truth. In contrast to this state of chaos, dàtóng represents a social order in which whole and part are one; the interests of the individual come to be inseparable from those of the fam-ily, village, and state that nourish and protect them. Within dàtóng, good governance is not experienced as a limit to desire but as desire's deepest fulfillment. For the practiced Confucian, doing what one wants is doing that which encourages the harmony and success—one might say the sustainability—of the community as a whole. This contrasts sharply with contemporary liberalism, which also makes a claim for the way in which private desire serves a broader public good. For classical liberals, self-interest is said to increase the efficiency and wealth of a society, but there is never any suggestion that it deepens spiritually the connection of an individual to his community, nor even that the latter as a whole is made more sustainable or just by virtue of unbridled selfishness. For Confucius, contemporary arguments about the incompatibility of individual freedom and happiness with strong ecological governance would have made little sense because both simultaneously extend and deepen the other. Life's every gesture—from the correct way to dispose of a fish bone to the manner in which one addresses a visiting friend—would be mediated for Confucius by right ritual, or *li* (禮). This does not imply a condition of mere repetition or even totalitarian control; rather, it indicates a system of reverential attention in which relations with the self, others, and nature are saturated with self-reflection and care. In dàtóng, no action is free from the burden of mattering. People in the ideal Confucian society are productively occupied but never overtaxed because the ruler encourages economic activity as a means to shared prosperity rather than unlimited private enrichment (as is in contemporary capitalist societies). From this perspective, prosperity is not an obstacle to but proof of moral rectitude. At the instant the former is gained at the expense of moral (or cosmic) truth, it becomes odious and toxic. It is important to stress that the image of Confucius as a dour Kantian moralist is a misguided Western projection: virtue, as in the philosophy of Baruch Spinoza, engenders overflowing joy, not unfeeling obedience. The ideal Confucian life is one lived in a state of continuous joyous attachment to learning, friendship, and music. Dàtóng does not posit a whole that lives on the back of strangled pleasure but rather replaces one system of pleasures with another. Today's most forward-thinking environmentalists—Vandana Shiva or Paul Bloom—imagine just such a project. The task is to rechannel pleasures in new directions rather than simply truncating them with a paternalistic "thou shalt not."[2]

In the light of Confucian utopia, late capitalism—the world we inhabit today—looks strangely empty. Precarious employment, shocking levels of inequality, a popular culture indistinguishable from advertising, widespread depression and anxiety, the emergence of violent xenophobic populisms, and the prospect of imminent ecological collapse all mark our era as a time of slow-motion ruination. Decay, rot, ruin: these terms would be the kind used by Confucius to understand our moment. Of course, these concepts presume an objective norm by which to measure the gap separating health from sickness. This is a way of thinking foreign to our own time, which tends to view such claims from within a common-sensical relativism that sees the individual as the highest arbiter of the true. So rich, however, is the image cast by dàtóng that it might be strategically necessary to suspend our usual poststructuralist suspicions and to pretend to believe (for an instant or two) in good old-fashioned social fullness. In place of a self-regulating liberal subject, dàtóng substitutes a social being rich in bonds as well as debt. Life, it suggests, requires ancestors, the grace of good harvests, and the continuous labor of farmers and mothers to exist. Even today, these seemingly outdated (or even folksy-cliché) elements hold true as the basic conditions of our collective being. With Aristotle (and against competitive individualists such as Thomas Hobbes), Confucius affirms a picture of existence as intrinsically obliged: society and its duties are not imposed onto the self; nor are they produced as an effect of a contract it can later choose to abjure or flee. The direct consequence of ignoring this indebtedness—of choosing the route of what Confucius calls the *xiaoren* (small man, 小人)—is a life without joy, peace of mind, and knowledge. This point is particularly valuable for the way it sets up the possibility of an immanent critique of capitalism—a critique that comes from within rather than from without. Society is not simply opposed to self-interest. Rather, it is an extremely narrow (liberal) conception of the latter in which the good life is equated with freely chosen pleasure, consistent rule of law, and the sovereign right to be left alone. The *xiaoren* thrives in cliques and gangs, forming a sense of right on the basis of local interest; the *junzi* (gentleman, 君子) takes the universal as the unshakeable foundation of moral judgment and strives to live a life compatible with the ideals of dàtóng. It would not occur to Confucius to provide a proof for the existence of others. Where Descartes establishes reality from the inside out, with certainty starting at the self and moving tentatively outward (toward the existence of others, society, nature, and so on), Confucius presumes from the beginning that there is a preexisting and structured reality of which humans constitute a limited yet significant part. Confucius, like Spinoza, sees in other people not barriers to the realization of the self—enemies on a territory of scarce pleasures—but as neighbors whose health, happiness, and justness makes possible a richer experience of selfhood.[3]

Furthermore, in contrast to an era in which profit seeking is the highest form of activity—and in which money is imbued by governments and entrepreneurs with godlike capacities to redeem humankind via the cliché of market innovation—the strictures introduced by Confucius make it impossible to transform living human labor into dead (and exploitative) economic capital. Wealth that takes the form of collectively shared abundance is celebrated by dàtóng, but it cannot be produced at the expense of the bodies of the poor; it is more equally distributed among those who produce it; and it cannot be allowed to distract from the far more important values of study, virtue, truth, and culture. Though Confucius does not outright reject accumulated individual wealth (or economic inequality)—a position he shares with Aristotle—he tends to celebrate the kind of frugality we more commonly associate with Plato's Socrates.

Most importantly, as an ideal, dàtóng constructs a specific form of political time, one that never permits the present to confuse itself with the end of history. Because liberal societies imagine themselves as constructed around unchangeable primary units—desiring individuals who in the last instance are necessarily selfish and whose tastes or pleasures are basically axiomatic—they tend to resist framing the future as the product of a collective intention. Liberal theory posits private vice as public virtue and does not see the individual as in need of a fundamental formation (or *Bildung*) beyond the basic technical training it needs to enter the economy. Though Confucianism measures the distance between here and utopia with an absolute standard—a *dao* with only one door—its belief in the perfectibility of people is a remarkable counterweight to the unrelenting realism of capitalism, a system that is forever confusing what is with all that has been or could be. In this sense, dàtóng occupies the same position within Confucian practice that communism does in the history of Marxism. It points to the possibility of collectively constructing something better.

We are no more capable of returning to the future dreamed of by Confucius than Confucius himself was his ancient Zhou utopia. Indeed, such a return is as undesirable as it is historically impossible. Dàtóng envisions a society in which parts are arranged in sets of relations— father and son, ruler and ruled—that are themselves fixed eternally by nature. It is this rigidity on the level of social infrastructure that creates the conditions for the emergence of a strong sense of moral, political, and aesthetic purpose. Moral plenitude is the other side of societal closure, a system the core logic of which is immutable and modeled on the fixed quality of the cosmos itself. In other words, moral truth of the kind proposed by Confucius presupposes a closed cosmos that it then aspires to imitate. Certainly the Western conception of Confucianism as a system of unthinking vertical enslavement or authoritarianism is exaggerated. Though a father should, in the last instance, be obeyed by a son, the ruler himself is held in check both by the tacit consent of the population (expressed in its willingness to be ruled) and transcendental moral norms. In contrast with the position of Hobbes, to overthrow an unjust king is to overthrow a common criminal or thug rather than a sovereign proper—which makes the death of a bad king an act of just punishment rather than open regicide.

There are genuinely egalitarian aspects to the Confucian position.

In the world envisioned by dàtóng, government would not be in the hands of a hereditary economic elite but would rather be run by a moral aristocracy of competent scholar-bureaucrats drawn (theoretically at least) from any class. This porousness, however, ends at the division separating male from female subjects: the latter are ascribed a set of ideal tasks and functions on the basis of their sex and barred without exception from participating in the Confucian bureaucracy. (In dàtóng, we are told, "men had their tasks, and women their hearths."[4]) The general form of government is likewise beyond critique: in dàtóng, the "will of the people" is restricted to affirming or denying the right to rule of a specific individual, but not the system that divides humans into rulers and ruled. All human behavior falls under the governance of norms of right practice that are universal. Harmony reigns in dàtóng because individual desire has been cultivated to delight entirely in the true unity of custom. Implicitly excluded are the alternative (and perhaps less orderly) delights of (for example) critical science, goalless wandering, or enticing, socially "useless" sex. The universality implied by dàtóng, its claim to safeguard the whole, is such that it cannot comprehend cultural difference except in the form of a deviation from right practice. Though it is theoretically open to others—Confucius claims that "barbarians" too can be improved by study—these can be included in the universality of dàtóng only on the condition that they shed their specificity and history. Dàtóng in this sense can be expanded to include the rest of the world, but only on the condition that the latter enter into universal (cultural) Chineseness.

Crucially absent from the concept of dàtóng is any real consideration of the relationship between humans and nature. Though Confucius sees in the ideal political order a reflection of cosmic or natural necessity (it should echo the *dao* of nature), he, like Socrates, is far more interested in humans and their ethical dilemmas than he is in either physical or metaphysical accounts of nature. Traditional Confucians are too busy looking at humans to notice the streams, mountains, and fields so beloved by Daoists and the great hermit painters and poets of the Chinese tradition. These issues come together in the *Book of Rites*, which frames the correct performance of ritual—be it the welcoming of a guest or the commemoration of a dead ancestor—as the line separating the human from the "merely" animal. After all, if the human does not separate itself from the animal via the attentive precision and self-consciousness of ritual, then what prevents mother and son, father and daughter from taking each other as mates? There is an undeniable speciesism at the heart of Confucius.

Dàtóng's limits do not stand in need of a knowing European update. Instead, it is to the Chinese intellectual tradition itself that we can turn for the resources needed to expand and repurpose a properly dialectical and ecological great universality. On the issue of the humanism of Confucius—his tendency to focus on intrahuman relations rather than those between humans and other forms of life (or nonlife)—the Confucian tradition itself begins to generate new sets of questions and answers toward the end of the Tang dynasty (AD 618–907). In part as a response to the more metaphysically curious Buddhist and Daoist positions, neo-Confucianism (or *daoxue*) begins to attend with a higher degree of specificity to what it claims is the overarching metaphysical structure of reality. This turn toward metaphysics—one that might be expected to marginalize empirical material reality by turning attention to ideal or transcendental structures—in fact worked to expand the frame of traditional Confucianism to include a more subtle account of the relations between human and nonhuman entities. Eighteenth-century European commentators interpreted Confucius as a protoempiricist Enlightenment figure, as an atheist with little to no interest in metaphysics; however, it is paradoxically by shedding precisely these empiricist leanings that the Confucian tradition reorients itself toward totality and in so doing the political terrain of the particular itself (things, animals, etc). In turning toward a new thought of the whole, relations between particular things (and the way we experience or politicize them) are changed.

In the concept of *qi* (氣), for example, the tradition begins to articulate the outlines of a material/psychical force or energy that flows through and binds the whole of human and nonhuman existence. Importantly, this gesture to deep ontology does not become the alibi for a repression of the particular but instead extends the reverential attention of ritual to include relations with individual nonhuman animals and objects.

Later Chinese thinkers, some of whom were non- or even anti-Confucian, would continue to expand and rearticulate the conceptual limits of dàtóng. In the late Qing period, conservative reformer Kang Youwei constructed a picture of dàtóng with explicitly liberal, socialist, and feminist elements. For Kang, dàtóng takes the form of a systematically envisioned utopia, a modern political future rather than an impressionistically idealized past. Though Kang was in practice a constitutional monarchist, advocating for reform within the terms of the existing imperial system, his theoretical work—published in full only after his death—called for a complete break with the logic of late Qing China. Where Confucius tethered dàtóng to the limits of the territorial imperial state, Kang called for the abolition of the war-prone system of nation-states and for its replacement by a single global order, one that would combine centralized political power with local forms of self-government. Where Confucius organized social functions according to the boundaries of sexual difference, Kang called for the abolition of inequality between men and women as well as all sex-based occupational divisions. Where Confucius remained a humanist, Kang saw the domination of animal life by human beings as an arbitrary limit to the reach of human *ren* (virtue); Confucius, he claimed, reduced love to a lazy (and destructive) "love-of-kind": it is precisely "because we only love one creature [ourselves]," he wrote, that we are not "averse to slaying all other creatures."[5]

One last point needs to be made here with respect the extremely important relationship between ideas and social reality. It is really not until the time of Mao Zedong that dàtóng ceased to be the name for a merely possible future and instead became the political objective of an organized mass politics. Without Mao's insistence on the need for a politics of struggle—on the existence of a future separated from its potential by actively entrenched forces and interests—dàtóng remains consigned to the saddest of fates: that of being a beautiful (but largely toothless) idea. This point is important because it separates an ethics of environmentalist care (for human and nonhuman life) from the difficult process of transforming this ethics into an effective institutional reality. To achieve the kind of universality aimed at by dàtóng, means and ends may not always entirely overlap. For example, such a struggle would almost certainly imply not granting equal pertinence or value to all opinions. Those with long-entrenched interests in the maintenance of ecodestructive capitalism could very well experience the universality of a green, politically organized dàtóng as dispossession, aggression, or exclusion. This is as it should be. Though Mao remains for many an unusably controversial political figure, it nevertheless remains the case that he knew well the difference between the avowal of an idea and the process by which it becomes a new form of practice. This is precisely the point of his famous quip about the difference between a revolution and a dinner party.

Provided we remain careful to acknowledge its own inherited limitations, the concept of dàtóng allows us to imagine a relationship to nonhuman life free of the reductive, humanist presuppositions of classical liberalism. Framed within an avowedly ecosocialist politics, it has the capacity to help displace contemporary associations of communism with the worst kind of extractivist industrialism. Mao's extremely important point, however, is that a concept (or a loanword, for that matter) is never de facto a politics. This is a point we must now still reckon with, recognizing that the domain of thought is a conflictual field, that new ideas immediately generate enemies, and that these enemies must be engaged politically if reality is to be changed for the better. The future is not a lost city, something infinitely behind (or beneath) us. It is not the purview of antiquarians and scholars (as Confucius himself thought). However, the future is also never simply something wholly new, a dream that has never been had before (a point made often by Hegel). Justice achieved, life collectively affirmed and effectively protected—this is at once the oldest, craziest, and most modern of dreams, and it is one we should never be too mature or informed to stop having.

Sueño ANOTHER PATH

NOTES

1. Confucius, *The Analects* (London: Penguin Books, 2014), 7.26.
2. See Vandana Shiva, "The End of Consumerism Is the Beginning of the Joy of Living," Ecowatch, December 21, 2012, https://www.ecowatch.com/; and Paul Bloom, "Natural Happiness," *New York Times Magazine*, April 15, 2009, https://www.nytimes.com/.
3. Confucius, *Analects*, 4.25.
4. William Theodore de Bary and Irene Bloom, *Sources of Chinese Tradition* (New York: Columbia University Press, 1999), 1:343.
5. K'ang Yu-wei, *Ta T'ung Shu: The One-World Philosophy of K'ang Yu-wei* (New York: Psychology Press, 2005), 265.

Pronunciation: fuːtminːɛ

Part of Speech: Noun

Provenance: Karen Ekman's *Blackwater* (1993)

Example: Protesters against the proposed supermarket and car park said that it would negatively impact the fotminne of long-term residents.

Our ways of remembering have changed considerably, especially with the technological advances of the last fifty years. Large-volume information storage and sharing services lend memories a disembodied quality; they float as if in a dematerialized cloud. There is a downside to the technological developments that let us broadcast our memories so easily and over such vast distance. Although online services enable us to record our impressions and distribute them to others across the globe, these records become detached from their original setting. As memories become separated from place, might humans lose the capacity for grounded ecological thought and action? The problem of humanity's detachment occurs at exactly the moment when our impact on the future of the planet is greater than ever before. Against these trends, the loanword fotminne, or "foot memory," reminds us of our primeval connections to the ground beneath our feet.

Collective memories are intricately dependent on the pathways we traverse and the physical places where we meet. The usefulness of the term fotminne rests on the assumption that whatever we do that

63

sustains memory also, and necessarily, sustains life. Fotminne was coined by celebrated Swedish novelist Kerstin Ekman. The term is Ekman's response to an increasing disconnect from what grounds our existence. In her novel *Blackwater* (1993), she brings to the crime genre her commitment to an ecological sensibility deeply rooted in her love for the heavily forested district of Jämtland in northern Sweden.

I recently went to Jämtland in search of firsthand experience of some of the places that have inspired Ekman's work. This is where the word fotminne evolved during the twenty years that Ekman lived in what the indigenous Sámi people affectionately call the Vaajma region (*vaajma* being the Sámi word for heart).[1] Situated directly in the center of Scandinavia, this region is home to Norwegians, Swedes, and Sámi. Its environment is characterized by the presence of overgrown wetlands called *myr* in Swedish, usually translated as "mire," though a close equivalent is also "fen"—an area of marshland, or former marshland, with few plant nutrients despite a neutral chemistry. A *myr* was once an Ice Age lake whose surface is now spongy with dense mosses such as sphagnum and other low ground covers. There are no paths on a *myr*, which means an inexperienced visitor can easily get lost. Furthermore, the ground here does not register human tracks. I encountered only the occasional imprint of an elk—the only creature heavy enough to leave its mark. In this context, I became increasingly aware of my lack of fotminne. It was impossible to walk along a straight line, but after the initial feeling of disorientation and unsettledness, it dawned on me that this was perhaps not such a bad thing. One walks slowly and cautiously on a *myr*, compared to the rush to reach a destination encouraged by straight roads that to Ekman epitomize our "death civilisation."[2] On this terrain, by contrast, one inevitably pays more attention to the ground, which Ekman describes in vitalist terms as a "network of paths, walking veins, memory vessels."[3]

Ekman's description raises questions about the ontological and material status of memory. How much does the soil's biodiversity contribute to human remembering? Does the biotic community, from microorganisms to plants and mosses each contributing their bit of biomass, act together to constitute a "being owing to which" memory is happening? Here I'm adapting Timothy Morton's formulation, "the being owing to which thinking is happening," which for Morton refers to an expansive preconscious (or protoconscious) hinterland from which consciousness emerges.[4] I don't claim to have any single answer to these questions, except to say I suspect that what humans recognize as memory is fed through "walking veins, memory vessels" by other-than-human contributions from the natural world. Precisely how much of a memory is owned by place is a moot point and would depend a great deal on the site itself, as well as how it is used, by how many people, and for how long. When I asked my guide, also a keen reader of Ekman, what the term fotminne evokes for him, part of his answer referred to communally held memories based on multiple visits by more than one person. The important point to make here is that the term fotminne goes a considerable way to remind us that both personal and cultural memory require a physical habitat to exist.

With the term fotminne, we get a sense of how place undergoes change through time as people, livestock, and wildlife traverse it. Ekman writes in *Blackwater* of the encounter between foot and ground as something happening presently in relation to both the long-term future and the recent past within the context of deep history. This nesting of timescales is also thematized in the overall structure of the novel since it is a murder mystery in which the chronological sequence of a crime unfolds in reverse as it is investigated. According to one critic, *Blackwater* is "one of the first environment-driven thrillers."[5] The novel is about a double murder that takes almost twenty years to solve, as the plotline moves back and forth between the years 1973 and 1991. Its original Swedish title translates as *The Events by the Water*, and locations, such as Blackwater, are often referred to as "events."[6] As such, *Blackwater* assembles an inventory of environmental and social changes that can be traced back to real-world events as reported anecdotally by the local press of the small village Valsjöbyn, where the author lived. This was brought home to me when my guide shared a family scrapbook of cuttings from the local press. This story of conflict between members of an intentional community and the local population dates to the 1970s, well before *Blackwater* was published. A similar series of events takes place in the novel, coupled with an explicit critique of specific manifestations of countercultural ideology. I realized that fotminne can also survive in stories after I had spent time learning my way in the *myr* and then contemplating this private archive.

In Ekman's writing, fotminne depends on an infrastructure of living and nonliving things. Nature becomes a character, and even rocks seem to shelter memory: "Remembering right out into the stony scree. Not getting lost. Remembering with your feet. Not with a sick tumour called longing which reproduces images wildly and crudely and crookedly. No, foot memories, leg memories. The capercaillie's coarse droppings—of pine needles, on pine needles—below a large pinetop he had ripped at with his beak."[7] If fotminne is "remembering with your feet," here Ekman gives the ground an active role by being solidly present, since fotminne is also the way in which "the ground responds to the foot."[8] Unlike representations that are often formed "crookedly," out of the distortions of anthropocentric desire, fotminne emphasizes materiality, as both a corrective to self-absorption and an appreciation of nonhuman contributions. Notice how in this passage Ekman draws attention to barely discernible traces "of pine needles, on pine needles," as though those "coarse droppings" of the capercaillie (a large forest bird) are an outgrowth of the physical world that makes the ground more visible. These signs become a record of creaturely life in the form of faintly readable traces.

Historic or natural settings help us preserve a sense of a collective as well as a personal past, but it would be a mistake to assume that the places themselves are like static bookmarks in our life stories. The term fotminne takes that notion further. In essence, fotminne means that places and paths are vulnerable to our passage over them. Distinct from the impressions we retain of where we have traversed, the places themselves are also inevitably impressed by our traffic, and this in turn affects their ability to serve as placeholders of our memories. Seeing this reciprocal relationship between how we comport ourselves and what will be preserved of our passage offers a new way to understand the now-ubiquitous term *sustainability*.

As an alternative to the global reach of cyberspace, fotminne dove-tails with literary critic Andreas Huyssen's emphasis on the local in what he calls "lived memory," all the more important in an age of information overload.[9] We are reminded that an active ecological sensibility depends on a selective and attentive form of remembering. The *minne* in the loanword fotminne comes from the Greek *mnēmonikos* (relating to memory) from *mnēmōn* (mindful). Unlike a term such as *carbon footprint*, which tracks carbon dioxide emissions quantitatively with an eye on their negative future consequences, what is affirmative about fotminne is its cautious hope for the positive impact of present actions. This is because the term draws attention not just to what is occurring right now but also to the impact of present actions on what will come to be remembered in the future. Given that *minne* is related to "mindfulness," the word might also be transliterated as "foot-minded." Fotminne looks forward as well as backward, referring to the ground's recollection of our passage that, once made, is retained in traces that can outlive individual lives. Our actions and practices affect our memories and how we will be remembered.

If in our walking around and experiencing the world we commit the ground to our memory, fotminne reminds us that each stroll or journey might strengthen a commitment between place and history, binding our future to that of the ground. What emerges in Ekman's articulation of the concept is that attention to our fotminne of a specific location contributes to our and that place's long-term survival. This, she reminds the reader, is particularly true for Sámi knowledge that fades from view as indigenous terms for and relationships with local flora and fauna are forgotten. The Sámi world itself withdraws when those ways of walking and naming that world cease to be practiced. With fot-minne, the notion of *habit* is allowed to resonate within *habitat,* so that routine and custom become wedded to the places they are practiced.

In *Blackwater*, Ekman draws attention to the practice of motherhood in the form of a precarious relationship between mother and daughter. The fragile nature of that relationship haunts the moment when the main character, Annie, warns her sleeping daughter that she is not an immovable place to which the daughter may in the future return at will: "I am not marked and demarcated. I happen. Mobility."[10] Here the mother speaks for what grounds all life; however, her tight-lipped and desiccated language suggests a source of life whose heart might grow stony over time. Like Annie, the earth is subject to change and is not always destined to retain the marks and demarcations that humans or any other species impress on its soil. What we may think of as lasting improvements can easily become the seeds of our destruction.

It would be wrong to assume that the word fotminne has relevance only for the few of us who can remain in and care for one area for pro-longed periods of time. For those who frequently travel, move house, or even relocate thousands of miles, including refugees, the cultivation of fotminne refers to our mindfulness of the sometimes dramatic and rapid changes that our traffic brings to the places we inhabit. The pro-liferation of human impacts that continually recreate the most heavily trafficked regions of the earth (popular tourist destinations, areas where economic development spurs accelerated urban growth, and so on) complicates the notion of ever returning to such places after spending time away. Geographer Doreen Massey theorizes that geographic places have lives of their own, distinct from the lives that pass over them.[11] Put simply, the locations where we arrive and depart are themselves moving through time. All sites that we traverse or inhabit, natural or man made, are in part constructed and maintained by our passage. However, despite their transience, these busy places do not inevitably outgrow the relationship they form with any specific traveler. The complex reciprocal relationship between movement and stasis that Massey describes is one of the clearest accounts of the reason why attention to fotminne is essential.

This contingency of place, as suggested by Massey and Ekman, illustrates the complexity of our individual and collective impact on where we live. If places themselves can be thought of as mobile or transitional, then this transience will be felt by those who reside in them as well as those who leave and hope later to return. This heightened sense of impermanence is an invitation to integrate human movements into long-term perspectives not limited by the temporal boundaries that circumscribe individual lives, nor ending when they end. Reflecting on our fotminne might therefore invoke ways of relating to the earth that span generations, even millennia.

Archaeological insights into a living oral tradition traceable to prehistoric Sámi culture in the far north of Scandinavia are relevant here. In her study of a Stone Age building style spanning 2,500 years, archaeologist Marianne Skandfer has found evidence of the extreme endurance of specific, place-based memories. Skandfer's research finds that a building style in use at the Finnmark coast from about 2000 BC appears to have had an inland reprise in the centuries around the time of Christ. She sees this as a sign that "a form of collective memory" prompted the hunter-gatherer society to revisit that building style.[12] Skandfer welcomes the stories and myths of the people now living in this area and includes them in her presentation of the archaeological record along with other scholarly research. She notes that these shared narratives teach the young how to orient themselves within the landscape "so that they learn to travel in it and use it, conducting themselves well, in order to take part in the previous generation's experiences."[13] This archeological evidence of past land use that guides each generation is fotminne, a knowledge of practical and ethical comportment kept alive through stories that conduct foot traffic along sustainable pathways.

Fotminne refers to the kind of practical knowledge Skandfer has identified. Survival of cultural traditions as well as the livelihood of Sámi practices depends on intergenerational remembering. Culture itself, as Ekman notes, is largely memory.[14] She points out that, as with an indigenous way of life such as that of the Sámi, human survival means preserving knowledge beyond what can be remembered by individuals. Fotminne refers to actions that are routine, repetitive, and everyday, and it imbues them with the measure and force of secular ritual. Seen as the basis of fotminne, to-and-fro traffic becomes a form of weaving that integrates memory, culture, and the environment. The practice of fotminne must be seen in contractual terms, like a handshake, as though the foot strikes a deal with the ground.

NOTES

1. The Sámi are indigenous people who have inhabited the northern regions of Scandinavia for around 5,000 years. They now live as seminomadic reindeer herders, though traditionally they also pursued coastal fishing and fur trapping.
2. Anna Paterson, "Landscapes Remembered: Kerstin Ekman and Nature," *World Literature Today* 82, no. 4 (2008): 45.
3. Kerstin Ekman, *Blackwater*, trans. Joan Tate (London: Vintage, 1996), 413.
4. Timothy Morton, *Dark Ecology: For a Logic of Future Coexistence* (New York: Columbia University Press, 2016), 24.
5. See Beverly Lowry, "Death in the Forest," *New York Times*, March 17, 1996, https://www.nytimes.com/.
6. Ekman, *Blackwater*, 301.
7. Ekman, *Blackwater*, 413.
8. Ekman, *Blackwater*, 414.
9. Andreas Huyssen, *Twilight Memories: Marking Time in a Culture of Amnesia* (New York: Routledge, 1995), 38.
10. Ekman, *Blackwater*, 278.

11. See Doreen Massey, "A Global Sense of Place," in *Space, Place, and Gender* (Malden, Mass.: Polity Press, 1994), 146–56.

12. Marianne Skandfer, "Ethics in the Landscape: Prehistoric Archaeology and Local Sámi Knowledge in Interior Finnmark, Northern Norway," *Arctic Anthropology* 46, no. 1/2 (2009): 97.

13. Skandfer, "Ethics," 92.

14. Jonas Gren, "Att Kliva av från Asfalten och Se: En Intervju med Kerstin Ekman," *10Tal Klimatsorg* 12, no. 13 (2013): 36–43, 40.

Ghurba

Allison Ford and Kari Marie Norgaard

Pronunciation: ghurbah (ʀurba)

Part of Speech: Noun

Provenance: Arabic

Example: Missing the familiar birdsong that no longer greeted me in the morning, I felt a sense of ghurba, remembering the tricolored blackbirds that lived in the nearby marshes before climate change had made the place inhospitable to them.

There is a familiarity to daily life that most of us relish. We go forward with a fair amount of confidence that tomorrow will resemble today. The details may differ, but the structure of our lives will persist. We will step outside of our door and see the same landscape. The same birds will greet us in the morning, and the same trees will line our street. Our family and friends will be where we left them, and we will follow the same habits and routines. This is especially true for those who have access to resources that buffer them from contingencies that life can bring—the class-privileged citizens of wealthy nations who are more resilient to disasters than marginalized people who struggle to meet their basic needs. In good times, daily life is familiar, and there is comfort in that. Of course, we don't really know what the world will be like in a year, let alone five or twenty years. But in the face of uncertainty, our ideas about the future are built on the shape of our present.

It takes only a few seconds of viewing the temperature graphs of climate forecasts to grasp that consistency in the experience of daily life may no longer be possible. Faced with unprecedented global environmental risk, our collective dreams of the good life threatens to come undone. In the age of climate change, the safety of some is predicated on the undoing of the daily lives of others. Social theorist Ulrick Beck calls this phase of modernity "risk society," a phase of capitalism marked by unprecedented social and environmental risk.[1] Even as industrial developments have made life safer for some people, industrialization has also led to a proliferation of new and unprecedented risks.

Risk society changes the social landscape. The established cultural values of progress, affluence, and what it means to be a good citizen of a wealthy, industrialized democracy will no longer hold in a climate-constrained world. This fundamentally challenges what Anthony Giddens calls ontological security—a sense of the continuity of everyday life.[2] Climate change portends the undoing of familiar ecological and social landscapes; it challenges the belief that the world of tomorrow will look anything like the world of today. Faced with global environmental risks, our sense of home as a place that is familiar and safe begins to come undone. To understand how to respond to such changes, we might look to the experiences of people who have already experienced the loss of the ordinary as a result of significant political, social, and environmental changes, such as Native Americans who have lost much of their historic homelands, displaced peoples such as Palestinians whose land is under Israeli occupation, and contemporary climate refugees whose homes are already disappearing around them. The experiences of people for whom risks have become realities might inform how we think about security in a changing world. What does it mean to be safe? What can we hold on to from our current lifeworlds? Where do we need to let go? How might we channel the collective desire for the good life into something creative—a force that propels us into a more equitable, sustainable future? First, let us name that desire. We borrow the word ghurba to describe a sense of longing for home that we apply to the ecological undoing of climate change, with all that we stand to lose.

Ghurba is an Arabic word that has no English equivalent. It translates loosely as "a longing for one's homeland." It evokes the deep connection that people have to their place of origin and the longing for place as something familiar and safe. Ghurba might be used to refer to the political longing of Palestinians cut off from their occupied historic homeland, or it might be used more casually by a student away from home for a temporary but prolonged period, missing home. Mohamed Awad, a scholar from Egypt, describes ghurba as "a sense of melancholy stemming from deep longing to a homeland, family, close ones, etc. It persists even if everything else in life is going well. It can also mean 'feeling lonely among people you know,' even at home. It's a state of mind just as much as it is a condition of physical separation."[3] Ghurba comes from the root word *gharb* (to go away, depart, or withdraw) and is linked to *gharib* (strange).[4] One departs and becomes a stranger, but the longing for familiarity endures.

In the age of anthropogenic climate change, we propose applying the word ghurba to name a longing for home that transcends place. We also want it to include the sense of connection to space, ecological and social relationships, and feelings of security that make up home. While we may associate home with a location, it is also imbued with cultural meanings that give shape to home as both a conceptual and physical space. Rising sea levels, ocean acidification, increased droughts and wildfires, and the loss of biodiversity, including culturally significant species such as salmon, mark the undoing of the ecological fabric that buttresses our daily lives. These changes force us to address the loss of ontological security that climate change entails. We can no longer assume that the world of tomorrow will be like the world of today. As weather patterns change, species move on to new habitats or die off, and increased storms, fires, and droughts roll in, we risk becoming strangers in places we thought we knew.

For many, climate change threatens to undermine the consistency of daily life that home provides, both conceptually and geographically. For example, the Karuk tribe along the Klamath River in Northern California has seen dramatic declines in salmon populations in their lifetimes. Recalling the environmental shifts that have also led to significant changes to social life, members of the tribe express feelings of loss that exemplify ghurba.[5] The decline in salmon changes social interactions, as well as how people internalize identity, social roles, and power structures. Karuk people describe grief, anger, shame, and hopelessness associated with environmental decline. Yet as Potawatomi scholar Kyle Powys Whyte emphasizes, indigenous people have endured many such shifts already as a result of colonialism, a process of violent disruption of their ecological practices.[6] Even if we stay rooted in the same place, the nature of that place may rapidly change around us.

Around the world, from the loss of land in island nations to the impacts of drought on the Syrian migration crisis, climatic changes are rendering homes unlivable. Since 2008, an average of 26.4 million people a year have been displaced by natural disasters—a number that is on the rise.[7] Some climate refugees may be moved preemptively to avoid disaster. In 2016, for example, the U.S. government funded the relocation of the entire community of the Isle de Jean Charles in Louisiana, which has experienced a 98 percent loss of land since 1955.[8] In other places, where the effects of climate change are more gradual or resources are not available to assist with relocation, the environment may change around people who cannot or do not want to leave, creating what ecocritic Rob Nixon calls "refugees in place"—populations subject to "displacement without moving."[9] Nixon includes climate change as a contributor to the "slow violence" of gradual and often invisible environmental damage that threatens to make a place unlivable. This is especially apparent in extreme cases, where entire homelands are being lost, as in the case of the Isle de Jean Charles. Chief Albert Naquin of the Biloxi–Chitimacha–Choctaw tribe was quoted lamenting the loss of his historic homeland: "We're going to lose all our heritage, all our culture." What was once the stuff of daily life is now "all going to be history."[10]

For already marginalized groups, the consequences of slow violence are especially severe. Indigenous peoples, people of color, women, gender-nonconforming folks, immigrants, the disabled, and others excluded or otherized by dominant institutions may be especially hard hit by the loss of home, which may serve as a conceptual space that offers security and well-being as a refuge against the powerlessness of oppressive forces. The experience of home is deeply shaped by gender, race, class, and other significant markers of social location. For example, Black feminist author and activist bell hooks writes that for people of color in predominantly racist American society, home can be a refuge against the vulnerability of violence and judgement. hooks describes the feeling of arriving at her grandmother's home after an uncomfortable passage through dangerous white neighborhoods where she regularly experienced racism: "Oh! that feeling of safety, of arrival, of homecoming when we finally reached the edges of her yard. . . . Such a contrast . . . this sweetness and the bitterness of that journey, that constant reminder of white power and control."[11] For hooks, home is where care and nurturing offered sustenance, development, and growth to those who were denied it in the public sphere. Although tinged with the melancholy of nostalgia, ghurba substantiates the valuation of home as a conceptual space that can protect against both symbolic and physical violence. Such a longing makes slow violence more evident by calling attention to what we are losing, even as it slips away.

As we link changing ecological circumstances to a sense of security in the world, we must consider how people in various social positions relate to the security of a home in the first place. In our respective research, we document the complex, often negative feelings that people have about climate change and its effects on daily life. Many people report feeling fear, anxiety, despair, and guilt at their part in producing greenhouse gases that are responsible for climate change, as well as helplessness about their ability to respond to a problem of such magnitude. These emotions are filtered and shaped by culturally specific feeling rules—the social norms that guide the expression of emotion in specific contexts.[12] Complex emotions about climate change must be managed; cultural norms shape forms of emotional management that make sense.[13] Climate change threatens to undo the fabric of our daily lives, to alter our access to life-sustaining resources and culturally valued relationships to place. We must manage the difficult emotions that arise as we feel less safe.

Emotions bind people to certain ways of being in the world. Unfortunately, for many people, the management of emotion translates into implicit denial, as people work to maintain normalcy by selectively ignoring important information, adhering to norms of interaction that prohibit serious conversation about climate change, or focusing the blame on others. For many privileged citizens of wealthy nations, their ability to continue their way of life depends on the denial of climate change. For example, one of the authors of this entry found that Norwegians, whose high quality of life is made possible by the exportation of massive quantities of oil, engage in collective processes of emotional management that allow them to maintain a belief in their own innocence, even as they know that the privilege of maintaining the continuity of a high quality of life is only made possible by the exploitation of poor nations and complicity in the generation of climate-changing greenhouse gasses. Given the attachment many of us have to consistency in daily life, what will it take for those who benefit the most from the practices that cause climate change to shift their relationship to emotions like powerlessness, fear, and guilt to acknowledge, address, and work through them, rather than let them drive us? As privileged people around the world are "faced with more and more opportunities to develop a 'moral imagination' and 'imagine the reality' of what is happening," will they adopt an "ecological imagination"—the ability to see the connections between human actions and their impacts on the environment—or will they continue to "construct their own innocence"?[14] Where business as usual predominates (which is almost everywhere within the wealthy industrialized world), citizens act to protect themselves from ghurba. Yet in the longing that ghurba evokes, might we find the seed of a different response?

Environmentalists and social scientists have puzzled over the missing link between knowledge about climate change and a commitment to adopting more sustainable habits. Accounting for the loss of ontological security that is threatened by the changes to our homelands that climate change portends may help explain the gap between awareness and action. Addressing the environmental consequences of our current way of life is a monumental task that threatens to undo our very sense of the continuity of daily life. For daily life to go on as normal, many people must maintain silence about the implications of their privilege. Adopting societywide practices that do not contribute to climate change may mean undoing many of the things we have come to rely on to make us feel safe, successful, and valuable. This is no small thing. Emotional attachment to our existing way of life is limiting our resolve to see clearly the changes we need to make to ensure the safety of our communities.

How do we acknowledge and accept the circumstances of profound risk in which we find ourselves without giving in to fear? People report experiencing complex, negative emotions about environmental risk, especially climate change, which can seem particularly abstract and hard to grasp.[15] Feelings like fear, anxiety, and despair can manifest in different ways depending on social context. Old ways of envisioning political modes of engagement have favored suppressing or minimizing emotions, framing them as weak or passive—barriers to rational, effective action. New ways of thinking about the role of emotions in social life challenge this, reminding us that emotions provide information about what is safe. Ghurba, the longing we feel for a homeland as a conceptual place of familiarity and safety, might point the way to an ecological imagination that will enable us to build future spaces that are inviting, sustainable, and safe.

In the face of change, many of us may be experiencing ghurba while lacking a name for the experience. It is hard, if not impossible, to transcend a nameless feeling. Failing to adequately process feelings of longing and loss may limit our ability to reintegrate changing circumstances into our lives. Ashlee Consulo Willox observes that grief and mourning are important aspects of coming to terms with climate change and its impacts, but that such emotional processes are silenced in public discourse.[16] Failure to adequately mourn may limit the possibility of future well-being as grief becomes stuck, blocking emotional processes. Mourning involves coming to terms with loss and reordering daily life in the absence of what has been lost. The first step, then, is to understand what is being lost. Naming ghurba as an element of our complex feelings about climate change acknowledges a longing for something other than what we have. Longing is a form of desire. And desire for a homeland, a community, and a relationship with the non-human world—beyond the alienated fossil-fueled consumer society we occupy—is the emotional power source of the work of building a more sustainable society.

In this historic moment of being confronted with deep collective loss, there is an opportunity to open ourselves to what comes beyond that loss. To do so, we must take stock of our history, our culture, and our politics to find the fragments that inspire ghurba. What is special about our homelands that we fear losing? How might we mourn those aspects that will be lost? How might we bring aspects of home into the future, so that when we arrive there we do not have to feel like strangers? Modifying our social institutions to mitigate and adapt to global risks like climate change will require change on all levels—in our political institutions, in our organizations, in our families, in our intimate interactions, in ourselves.[17] If we are to create a future that is equitable as well as sustainable, our environmental work will need to overlap with the work of social justice movements that demands a broader form of security for all populations, not just the privileged few who can afford it. The depth of transformation this calls for is potentially painful, but to ignore the imperative to change offers no protection.

How do we manage this tension between the anxiety of the contemporary political moment and the residual desire for some form of the good life? If we give up some of the convenience, disposability, and novelty of consumer life, what is left? A drive for materially safe, socially connective, environmentally secure lifeworlds that might keep us connected to the homes we may have to leave behind.

Solastalgia
ANOTHER PATH

NOTES

When something is borrowed, thanks are always warranted, and we are grateful to the Arabic-speaking world for this evocative word, and the native speakers who helped us understand its richness. We decided to borrow the word ghurba before the 2016 election, without knowing how extreme anti-Muslim and anti-Arab sentiment would get in Trump's America. In the current political climate, we consider it especially important to acknowledge our gratitude, and to honor the beauty and richness of Arab cultures that are so often overlooked in its cultural othering.

1. Ulrich Beck, *Risk Society: Towards a New Modernity*, ed. Mark Ritter (Los Angeles, Calif.: Sage, 1992).

2. Anthony Giddens, *Modernity and Self-Identity: Self and Society in the Late Modern Age* (Cambridge: Polity Press, 1991), 36.

3. Mohammed Awad, personal communication, 2017.

4. Hans Wehr, *The Hans Wehr Dictionary of Modern Written Arabic* (Wiesbaden, Germany: Otto Harrassowitz, 1993).

5. Kari Norgaard and Ron Reed, "Emotional Impacts of Environmental Decline: What Can Attention to Native Cosmologies Teach Sociology about Race, Emotions and Environmental Justice," *Theory and Society* 46, no. 6 (2017).

6. Kyle Powys Whyte, "Our Ancestors' Dystopia Now: Indigenous Conservation and the Anthropocene," in *Routledge Companion to the Environmental Humanities*, ed. Ursula Heise, Jon Christensen, and Michelle Niemann (New York: Routledge, 2017).

7. Michelle Yonetoni, "Global Estimates 2015 People Displaced by Disaster," Internal Displacement Monitoring Centre (Geneva, Switzerland), 2015, http://www.internal-displacement.org/.

8. Coral Davenport and Campbell Robertson, "Resettling the First American 'Climate Refugees,'" *New York Times,* May 3, 2016, https://www.nytimes.com/.

9. Rob Nixon, *Slow Violence and the Environmentalism of the Poor* (Cambridge, Mass.: Harvard University Press, 2011), 17.

10. Davenport and Robertson, "Resettling."

11. bell hooks, "Homeplace: A Site of Resistance," in *Yearning: Race, Gender, and Cultural Politics* (New York: Routledge, 1990).

12. Arlie Russell Hochschild, "Emotion Work, Feeling Rules, and Social Structure," *American Journal of Sociology* 85, no. 3 (1979).

13. Kari Marie Norgaard, *Living in Denial: Climate Change, Emotions, and Everyday Life* (Cambridge, Mass.: MIT Press, 2011).

14. Kari Marie Norgaard, "Climate Change Is a Social Issue," *Chronicle of Higher Education,* January 2016, https://www.chronicle.com/. On the construction of innocence, see Kari Marie Norgaard, "'People Want to Protect Themselves a Little Bit': Emotions, Denial, and Social Movement Nonparticipation," *Sociological Inquiry* 76, no. 3 (2006).

15. See Norgaard, *Living in Denial* and "Climate Change"; and Allison Ford, "The Emotional Landscape of Risk: Self-Sufficiency Movements and the Environment" (master's thesis, University of Oregon, 2014).

16. Ashlee Cunsolo Willox, "Climate Change as the Work of Mourning," *Ethics and the Environment* 17, no. 2 (2012): 141.

17. Matthew Schneider-Mayerson makes a similar point in "Affect," his entry in *Fueling Culture: 101 Words for Energy and Environment,* ed. Imre Szeman, Jennifer Wenzel, and Patricia Yaeger (New York: Fordham University Press, 2017), 28–30.

Godhuli
Malcolm Sen

Pronunciation: go-dhu-lee (goḍʰuli)

Part of Speech: Noun

Provenance: Bengali and Hindi

Example: As the day nears its end, at this godhuli hour of the planet, our bodies yearn for refuge.

The human body is akin to a heliographic plate, reflecting and absorbing sunlight, which transforms us at a cellular level every day. Nicéphore Niépce, the French inventor of photography, wrote to Louis Daguerre in 1829, "In the process of composing and decomposing, light acts chemically on our bodies. It is absorbed, it combines with them and communicates new properties to them."[1] In 1945, 116 years later, when the first nuclear bomb was detonated, humanity added another source of luminescence that would similarly affect our bodies. It was light like no one had ever seen before, and onlookers found their vocabularies severely tested when they attempted to describe it. The deputy commander of the Manhattan Project, Thomas Farrell, notes, "The whole country was lighted by a searing light with the intensity many times that of the midday sun. . . . It was golden, purple, violet, gray, and blue. It lighted every peak, crevasse, and ridge of the nearby mountain range with a clarity and beauty that cannot be described but must be seen to be imagined. It was that beauty the great poets dream about but describe most poorly and inadequately."[2] Seventy-one years after that first nuclear test, a group of geologists assembled at the International Geological Congress in Cape

Town, South Africa, and declared that the planet had entered a new geo-logical epoch, the Anthropocene. Stratigraphers point to the geological evidence left by nuclear radiation on earth to support the Anthropo-cene thesis. The Anthropocene acknowledges these new stratigraphic signatures—the presence of the indelible textuality of particulate mat-ter, vaporous gases, and radionuclides—that we have etched onto the planetary body. Increasingly scientists find that such stratigraphic sig-natures have corresponding somatic inscriptions as well.[3] The Anthro-pocene, according to some suggestions, properly begins in the 1950s, its genealogy illuminated by the unnatural "golden, purple, violet, gray, and blue" light of a nuclear detonation. The nuclearization of the planet is a good reason to expand our understanding of the human body as a heliographic plate. We are, as Elizabeth DeLoughrey so poignantly writes, "creatures constituted by radiation, solar and otherwise."[4] The same is true of the planet.

Sunlight takes only about eight minutes to reach the earth; its gifts of light and life are bestowed to us from the past. As we teeter on the precipice ushered in by global climate change, when our changing ecol-ogies violently shape us from without, and as we hesitate at this junc-ture of the Anthropocene, where microparticles and subatomic detritus shape us from within, how might we imagine a future which simulta-neously acknowledges its indebtedness to such past luminescence as well as its debris-saturated present? Godhuli (গোধূলী), a Bengali word, is a remarkable portmanteau. It refers to the time of day we might oth-erwise call twilight in English. However, it resonates with an ethics of place and a metaphysics of possibility in a way that the English word does not. Like twilight, when the sun is below the horizon and sunlight is refracted through the atmosphere, godhuli also refers to the fleeting moments that immediately follow sunset. But unlike twilight, godhuli's refracted light is located terrestrially, on an earthly plane rather than in the atmosphere. This is because, in Bengali, go refers to cows, dhuli to dust. Godhuli is thus the time of day when cows, with their hooves kicking up dust, return from pasture to their nightly refuge. The time of the day is so named because of the unique color caused by the conditions of light and dust. This word commingles light and dust, but also, importantly, color and texture. It is a word pigmented by the iron particulates and the burnt sienna of sunbaked Indian soil. Godhuli speaks of the rusty orange dust emanating from the earth as it responds to the hooves of cows seeking shelter. In sum, godhuli is more than its parts. In it, space and time, light and dust converge.

Dust may not initially seem like an obvious subject for philosophy or poetry. Its matter-of-factness and its ubiquitous presence in our lives seemingly robs it of any special notice. But the banal is often worthy of further deliberation. When dust appears as a motif in literature, it does so predominantly in negative terms: as matter that hinders and disorients. In *Heat and Dust*, Ruth Pawar Jhabvala's 1975 novel, dust obscures not only sight but also one's ability to think: "Dust storms have started blowing all day, all night. Hot winds whistle columns of dust out of the desert into the town; the air is choked with dust and so are all one's senses."[5] In James Joyce's "Eveline," dust is a metonymic signifier of the repetitive meaningless acts which add to the ennui that pervades Dublin, a colonized city: "Home! She looked round the room, reviewing all its familiar objects which she had dusted once a week for so many years, wondering where on earth all the dust came from."[6] Yet for all the reasons that make dust signify boredom and obscurity, without it, life as we have all come to understand it would cease to exist. This fact is not lost on Alfred Russel Wallace, the nineteenth-century naturalist and anthropologist who wrote poetically about dust. In 1898, Wallace noted that, like dirt, dust is seen as "matter in the wrong place."[7] But he argues that it is dust that makes the sky appear blue and avoids a "perpetual glare" of sunshine that might otherwise erode earthly life. "Without dust the sky would appear absolutely black, and the stars would be visible even at noonday. The sky itself would therefore give us no light. We should have bright glaring sunlight or intensely dark

shadows, with hardly any half-tones."[8] Nephology, the branch of meteo-
rology that studies clouds, confirms that we would also not have clouds
without dust; water droplets need particulate matter to cling on to as
they condense to form clouds. It naturally follows that there would be
no rain without dust. In this manner, the universal logic of harmonious
coexistence of antonymic entities is brought to fruition: rain could not
occur without that which it washes away.

Beyond its paean to dust, to "half-tones" of light and its acknowledgment of the passing of time, godhuli conjures up a number of expansive
associations in its original language. In the literary and musical works
of Rabindranath Tagore, it is often a metonym for a romantic or pen-
sive mood. In Tagore's song "Aji Godhuli Lagone" (Today at the godhuli
hour), the speaker awaits the arrival of his beloved, but he feels as if the
forests and the earth itself know much more about his beloved and her
thoughts than he does.[9] The natural world conspires and rejoices in its
secret knowledge; it seems that all the speaker can do is repeat his faith
in the beloved's imminent arrival, which forms the refrain of the song's
lyric. Godhuli, in Tagore's writing, is as much a temporal setting as it
is a cultural indicator of the romanticism and nostalgia begotten of the
captivating iridescence of a tropical twilight. But it is also more than
that. Tagore's lyric articulates through the registers of dust and light an
understanding of temporality that doubles as potentiality. Marking time
using the word godhuli underlines a notion of futurity as embedded
potential in the present. The contemporary is energized as a moment of
capacious possibilities in which Tagore's speaker's beloved may arrive.

In Hindu scriptures, godhuli also opens up a portal of potential
and demands human participation and action. In this way, reframing
twilight through this word recognizes the capacity of words from cultures we do not call our own to readjust our place in the world. In the
broader scheme of things, loanwords are remarkable tools to rethink
our present. What better time to multiply our vocabularies than in this
godhuli hour of the planet, at this very moment when we have just

recognized that we reside within a geological column (correctly, but also ominously) named after our species?[10]

That language shapes our perceptive capacities has been a matter of deliberation for some time. It seems that language determines something as basic as our conceptualization of physical phenomena, like light. German writer Johann Wolfgang von Goethe challenged Isaac Newton's understanding of the color spectrum (red, orange, yellow, green, blue, indigo, violet) on the basis of Greek perception of color. It would seem that to see the shade of blue that the Greeks saw, we also need to understand their aesthetics of color. Historian Maria Michela Sassi notes that Goethe claimed "that light is the most simple and homogeneous substance, and the variety of colors arise at the edges where dark and light meet. Goethe set the Greeks' approach to color against Newton's for their having caught the subjective side of color perception."[11] Sassi concludes that Goethe was right in challenging the "mathematical abstractions of Newton's optics." To see the world through Greek eyes, it is imperative to understand the Greek theory of color, without which we fail to recognize the importance of light and brightness in their "chromatic vision," and we cannot understand the "mobility and fluidity of their chromatic vocabulary."[12]

Sassi's work is a timely reminder to revise our understanding and perception of the world, which is increasingly mediated by the monochromatic world of a "universal" language such as English. It reminds us what Frantz Fanon wrote in 1952: "To speak a language is to take on a world, a culture."[13] While learning languages such as English or French offered colonized peoples of the nineteenth and twentieth centuries a passport into the imperial cultures of the West, such learning also increasingly distanced them from the vocabularies that allowed subjective and culturally relevant interpretations of their environments. Loanwords like godhuli offer a perceptive vantage point in the Anthropocene by allowing us to reconceptualize our planetary refuge and to rethink our contemporary moment as a time that demands such imaginative acts. If our aesthetic experience and chromatic perception of color may be culturally modulated, then our understanding of our place and time in the world may similarly depend on the languages we use—thus godhuli and its unique prism of light and dust.

Granular and particulate matter, like dust, have a special bearing on the Anthropocene. This is not simply because microscopic structures like atomic subparticles and microplastics haunt our everyday lives and bodies but also because the particulate allows us to understand the origins of human dominance over the planet. The beginning of the Holocene, approximately twelve thousand years ago, closely corresponds to what paleontologists call the Neolithic demographic transition (NDT). This was a time when our Neolithic ancestors moved from foraging and hunting to agriculture. By doing so, they unwittingly initiated a process that resulted in the first sharp increase in the world's population, which in turn began to alter existing planetary systems. The NDT marks the originating moment of a trajectory that ultimately leads us to the Anthropocene, to a time when humans recognize themselves as, to use historian Dipesh Chakrabarty's phrase, not simply actors of history but agents of geology, transforming and altering (this time knowingly) fundamental planetary processes.[14] The NDT does not "cause" the Anthropocene, to be sure; however, it speaks to us through a vocabulary of the granular and the particulate—that is, through seeds, kernels of concentrated potential, which made possible the unfolding of our history. Interestingly, through its visual economy of domesticated cattle, the human history of settled agriculture is encoded in the word godhuli. Godhuli's charisma relies on the dust (particulate matter) kicked up by the hooves of domesticated cattle, drivers of preindustrial agriculture. The word echoes a civilizational shift of which we are all inheritors.

The Anthropocene is a site of inherent contradictions. For example, the Age of Human names a species and its time on earth, but it also implicitly denotes that this epoch will come to a close at a future stage. It thus evokes the possible extinction of the human species and encodes a critique of human overreach even as it names an entire geological epoch after the human species. This contradiction should steer us to acknowledge the contemplative, imaginative, and empathetic paucities of our species. Some of these limitations may be the product of our neurological makeup but are definitively a result of the great inequalities built into the late modern, late capitalist systems that define our lived reality.[15] The seemingly impossible task of living a life outside of the capitalist system of interminable economic growth and production, consumption, and refuse radically skews our perception of the present, foreclosing a meaningful relationship with the deep past and the near future. In this temporal scheme, the past is something to be surmounted rather than negotiated, and the habitability of the future is never in question. Indeed, capitalist worldviews often project the future as already within reach. Consider, for example, the language that car manufacturers use to peddle their products. The tagline for the Audi RSQ will suffice here: "Tomorrow has arrived today."[16] Such temporalities are at odds with the kind of planetary future that is being shaped by anthropogenic climate change.

A new lexicon for the Anthropocene calls for a worldview that corrects our skewed perception of time. For this we need an alternative history to locate ourselves in the present, and we similarly need an alternative vocabulary to imagine the present as embedded futurity. "Hold to the now, the here, through which all future plunges to the past," thinks Stephen Dedalus in James Joyce's *Ulysses*.[17] This conception of time imagines the present as being imbued with the weight of the past and envisions it as an architect of the future. In this regard, the kind of secular and aesthetic interpretations of godhuli that I have been charting above finds meaningful resonances in the term's religious heritage. It should not be surprising that the allure of godhuli's optics, the slant of light rays traveling through dust that it translates into language, makes it a particularly auspicious time in Hinduism. Between the glare of the midday sun and the darkness of a rural night, godhuli, for Hindus, presents a window of opportunity for *karya* (work, action, or ceremonies, such as weddings). It offers a moment of opportunity to be made use of in the face of astrological hurdles and planetary maleficence. In Hinduism, godhuli thus offers a moment of hope, a way out when no other paths seem passable. At a time when we name the Anthropocene and recognize the potentiality of our threshold, our civilizational godhuli hour, a time for action and illumination, not least of which should be the recognition, as Nietzsche would have it, of our false idols. Godhuli enlivens the threshold on which life depends. Its call for a temporality that is determined by the "half-tones" conjured up by dust acknowledges at once the "sweet spot," the "Goldilocks zone" (terms used by scientists to denote the unique position of our planet in the solar system), in which the orbital path of earth is mapped. In the Anthropocene, godhuli energizes the recognition of the telluric origins of human history. The identification of the harmony between the particulate and the planetary is a necessary condition for meaningful life in the Anthropocene

Robert MacFarlane describes his book *Landmarks* as "a word hoard of the astonishing lexis for landscape." McFarlane's book is a stupendous meditation on the power of language to reassert our relationship with our environment. He writes, "We need now, urgently, a Counter Desecration Phrasebook that would comprehend the world—a glossary of enchantment for the whole earth, which would allow nature to talk back, and helped us to listen."[18] If the first nuclear detonation immediately called to mind the paucity of language to represent a reality as awe-inspiring as the mushroom cloud, then the Anthropocene similarly asks us to fashion a new language and expand our vocabulary to bear the weight of our contemporary moment. Often that might mean traveling beyond the monochromatic worlds of our mother tongues and in turn enriching our conceptual capacities to imagine, comprehend, and empathize not only with distant spaces but also with distant times. We inhabit an unprecedented threshold moment in the history of our species, and we need words that reveal that sense of in-betweenness. Godhuli illuminates that fact.

In the end, what is most enticing about this word, godhuli, is that it captures the luminescence that underwrites our physical existence on this planet. Like the cattle traversing an Indian field, it reasserts our relationship with and need for a place of refuge. Donna Haraway notes that "our job is to make the Anthropocene as short/thin as possible and to cultivate with each other in every way imaginable epochs to come that can replenish refuge. Right now, the earth is full of refugees, human and not, without refuge."[19] It is little wonder that at the heart of Haraway's observation is a call for an act of imagination. A loanword such as godhuli allows us an opportunity to make conceptual leaps, to take long views of history, and to better orient ourselves in this new geological epoch. Godhuli is not simply a paean to dust or a romantic fetishization of a chromatic wonder. It also reminds us of the planetary nature of our place of refuge and underlines the fact that we inhabit a time that demands imaginative labor to sustain that sense of homeliness.

Nakaiy
ANOTHER PATH

NOTES

1. Paul Virilio, *The Vision Machine* (Bloomington: Indiana University Press, 1994), 19.
2. Cited in Alex Wellerstein, "The First Light of Trinity," *New Yorker*, July 16, 2015, http://www.newyorker.com/.
3. Ever since the first nuclear tests carried out in the United States, medical practitioners have studied resulting changes in the human body—for example, in the chemical composition of bone structures. One of the earliest of such studies was on concentrations of strontium-90 in baby teeth. It was carried out in St. Louis in 1959. See "Teeth to Measure Fall-Out," *New York Times*, March 19, 1959, https://timesmachine.nytimes.com/timesmachine/1959/03/19/89163843.html?zoom=14.85&pageNumber=67.
4. Elizabeth DeLoughrey, "Radiation Ecologies and the Wars of Light," *Modern Fiction Studies* 55, no. 3 (2009): 468.
5. Ruth Pawar Jhabvala, *Heat and Dust* (London: Hachette, 2011).
6. James Joyce, "Eveline," in *Dubliners* (Oxford: Oxford University Press, 2000), 25.

7. Alfred Russel Wallace, *The Wonderful Century: Its Successes and Its Failures* (Toronto: George Morang, 1898), 69.

8. Wallace, *Wonderful Century*, 81.

9. Rabindranath Tagore, "Aji Godhuli Lagone," *Swarabitan* 58: n.p. The song was originally written in 1937.

10. A geological column is a system of classification for the layers of rocks and fossils that form the earth's crust.

11. Maria Michela Sassi, "The Sea Was Never Blue," *Aeon*, July 31, 2017, https://aeon.co/.

12. Sassi, "Sea Was Never Blue."

13. Cited by Bill Ashcroft, *Postcolonial Transformation* (London: Routledge, 2013), 57.

14. Dipesh Chakrabarty, "The Climate of History: Four Theses," *Critical Inquiry* 35, no. 2 (2009): 206.

15. See, for example, Daniel Kahneman, *Thinking Fast and Slow* (New York: Farrar, Straus & Giroux, 2011); Jason Moore, "Ecology, Capital, and the Nature of Our Times: Accumulation and Crisis in the Capitalist World-Ecology," *Journal of World-Systems Analysis* 17, no. 1 (2011): 108–47.

16. AdForum, Audi, "I, Robot 2," https://www.adforum.com/creative-work/ad/player/52190/i-robot-2/audi.

17. James Joyce, *Ulysses*, Gabler edition (1922; reprint, London: Random House, 1986), 153.

18. Robert MacFarlane, "Desecration Phrasebook: A Litany for the Anthropocene," *New Scientist*, December 15, 2015, https://www.newscientist.com/.

19. Donna Haraway, "Anthropocene, Capitalocene, Plantationocene, Chthulucene: Making Kin," *Environmental Humanities* 6 (2015): 160.

Gyebale

Jennifer Lee Johnson

Pronunciation: je-baa-leh (ʒɛːbʌːlɛ)

Part of Speech: Salutation

Provenance: Luganda

Example: Upon seeing a neighbor replacing their formerly well-manicured lawn with native prairie grasses, a stranger collecting aluminum cans to recycle for a bit of cash, or a protester blocking traffic to halt the construction of an oil pipeline, you might say, "Gyebale!" Your interlocutor might reply, "Gyebale!"

When two adults meet in passing along the southern shores of Uganda, they do not say hello. Instead, they say *gyebaleko*: thank you for the work you do. *Gyebaleko*, or more colloquially gyebale, is an informal greeting used to acknowledge the presence of others by first expressing appreciation for the contributions they make toward the everyday work of living well together. It is a common courtesy extended no matter how large or small a person's work may be, whether it is paid or unpaid, known or unknown, or ongoing, already completed, or to be started sometime in the future. When one is greeted with gyebale, the customary response is to return the sentiment with *kale, nawe,* gyebale—literally, yes, and you, thank you for the work *you* do. Regularly recognizing the work of others, and being recognized for your own, strengthens relationships required to live and work well into increasingly uncertain futures.

As my Ugandan colleague and self-proclaimed fisherman by birth

Bakaaki Robert explains, everyone—young and old, rich and poor—will do some kind of work worth doing on any given day. There may be children, gardens, livestock, and businesses to tend, fish to catch, dry, and sell, water to carry, grounds to sweep, clothes to wash, meals to prepare, books to study, homes to build, buses to catch, and almost always money that must be earned. Gyebale is a word quickly but generously uttered, so that the everyday work of living well with others may continue, no matter the social standing of an individual or the specific nature of the work they do.

Greetings are important. In Uganda, as in many other cultural contexts, it is exceedingly rude to encounter another person without verbally acknowledging their presence. Family, friends, acquaintances, and strangers all deserve the mutual respect that convivial recognition affords. Failing to greet another properly may engender suspicion, mistrust, or ill will between parties. The regular practice of exchanging greetings helps people to remember past social interactions as they negotiate present and future ones. For those who say gyebale, ignoring the presence of others is no way to lead a life worth living.

Gyebale is most commonly used as a brief informal greeting when either party appears unable to engage in lengthy formal greetings. Indeed, it is often impolite to expect individuals actively hard at work—for example, carrying heavy things, heading quickly from place to place, or already engaged in close conversation with another—to begin extended routinized greetings. In these moments, gyebale offers a conscientious way to extend one's appreciation and encouragement for the ongoing work of another within shared contexts without having to impede their activities (or have them impede one's own). The appropriate time to greet another with gyebale is not unlike wishing someone happy birthday. The good wishes are welcome any time—before, during, and after the event itself. In Uganda, those who spend their days in freshly pressed white-collared shirts behind computers in air-conditioned offices and those who wear tattered T-shirts and trousers to dig and sling dirt in the midday sun all may be thanked for their work and thank others with gyebale. All people in all walks of life may use gyebale, but it is an especially useful greeting to offer friends and acquaintances actively engaged in work, and for visitors, hosts, and travelers of all kinds.

Gyebale belongs to a category of greetings common within a wide geographic and linguistic range of African languages, but without a clear equivalent in American or British English.[1] Although there are many greetings within African vernaculars specific to particular kinds of activities, most address others at work.[2] By way of a few examples: for speakers of Kerebe on Ukerewe Island in Tanzania, the greeting is issued as a question, *Milimo?* (How is work?), with the person at work responding *Milimo nizyo!* (Yes, this is work!), meaning "the work is going on well."[3] In Kinyakusa, also spoken in Tanzania, the term is instead *ubombile* (you have worked).[4] In Igbo, spoken in southeastern Nigeria, it is *daalu olu* (thank you for work/greetings to you at work) or *onye oly daalu* (worker, thank you).[5] A number of additional African greetings, including gyebale, are glossed specifically in English as "well done."[6] For example, in Uganda, the vernacular term [o]gyebale used in the Runyoro, Rutooro, Luganda, and Lusoga languages and *apowyo* used in Dholuo are all glossed in Ugandan English as "well done."[7] The Yoruba phrase *e ku ise*, according to Femi Akindele, and the Akan phrase *mo ne adwuma*, according to Kofi Agyekum, are similarly glossed as "well done" in Nigerian and Ghanaian English, respectively.[8]

Like gyebale, the African English gloss of "well done" has no clear Euro-American English equivalent. In Euro-American contexts, the use of "well done" reflects high praise after an evaluation of exceptional work already completed, whereas gyebale does its work more as a verbal tip of the hat to others in passing. Because much of the work that needs doing on any given day may be considered mundane, routine, or beneath those with comparatively higher social standing in Euro-

American contexts, much of the necessary but unglamorous work that gets done every day often goes unacknowledged. Those who greet each other with gyebale (or *e ku ise, apowyo, mo ne adwuma*, and many others) use "well done" to reflect appreciation for everyone's work, which is always in progress. Rather than a superlative salutation offered after exceptional achievement, in African English, "well done" is used in affirmation of one's existence and importance in relation to others, as well as the contributions of others toward more collectively distributed forms of well-being. In the process, everyday and exceptional work of all kinds is encouraged, appreciated, and sustained.

Gyebale is notably distinct from similar words conveying thanks in use in Uganda. For example, *webale*, very similar to the English phrase "thank you," is used to extend appreciation to another for a specific thing or act that directly benefits the well-being of the receiver. For example, when I enjoy a hot meal shared with a Ugandan friend who prepared the food, I say to her *webale okufumba* (thank you for cooking). Rather than thank me for eating, she would likely reply *webale kusima* (thank you for appreciating). Unlike "thank you," which connotes gratitude for a specific act done for a specific person or group of people, gyebale is used to express appreciation for work that may not be immediately apparent but nevertheless generates collective benefits.

Although gyebale is often uttered informally, it is sometimes used to begin or signal an end to extended formal greetings practiced in accordance with the situated norms of speaking and being spoken to. The proper practice of formal greetings in Uganda, as elsewhere, are shaped by age, gender, and the particularities of any given meeting. While participants may make and answer inquiries about one's night, day, health, work, family, and whatever and whomever else they find it appropriate to discuss, it is common to shake and hold hands. Depending on the social standing of each greeter, one or more participants may even kneel and respectfully avert their eyes as a sign of deference. One's ability to greet others properly (or not) demonstrates the quality of a person's upbringing and their awareness of the importance of their relationships with others within a shifting and sometimes fraught social landscape.

Indeed, it is within contexts where extended greetings are often highly ritualized, required, and actively interpreted that the use of gyebale in passing does its own work. In this most frequently used variant of the term, gyebale offers an informal and universally appropriate greeting devoid of established social hierarchy, implied evaluation, or impossible expectation.[9] Those who continue to make themselves known to others in passing with gyebale work—intentionally or otherwise—to establish common convivial ground for extended and sometimes difficult conversations in the future.

Ugandans certainly recognize that all work is not necessarily good work, and that work that some consider good may hinder the ability of others to do work they deem worth doing. This has particularly been the case in recent years, when the Ugandan government has increasingly harnessed the military might of the state to limit public protests led by opposition party candidates and to dispossess residents of their land to facilitate resource extraction, expand highways, and clear formerly dense forests to seed palm oil plantations. Still, those who labor toward projects big and small, projects good, bad, or morally ambiguous, say and are told gyebale. Whether said in passing or as part of formal greetings, gyebale creates opportunities for continued work and also establishes a basis for mutual recognition from which to continue conversations into the future. Although there may be many reasons not to thank a member of the military police for quelling a potential riot, by failing to say gyebale in moments when it is expected and appreciated, conversations about the potentially harmful work of another may never begin.

After working gyebale into my lexicon and habits in Uganda, I came home to a cold winter in the Midwestern United States to find my American English lacking a satisfying equivalent. There seemed to be no simultaneously simple and meaningful way to acknowledge the work of a woman sprinkling salt on a public sidewalk so that passersby did not slip on the ice; there was nothing appropriately quick and unobtrusive I could say to a man patiently answering his young daughter's questions in a busy grocery store checkout line. There was nothing but gyebale I could say to the team of men hauling trash from my neighborhood's dumpsters away and out of sight. In those moments, when I longed to say something, I said nothing at all.

Acknowledging the good work of others may well offer vital support for the everyday acts and arts of living and working well into increasingly uncertain futures. Although I have yet to hear Ugandans saying gyebale to other-than-human beings, they do express gratitude through interspecies comparisons. To give just one proverbial example: when you eat flying ants (or termites, a delicious and nutritious seasonal delicacy), don't talk about them being tasty without appreciating the hard work of the *nkuyege* (work-termites) that have built up the anthill.[10] For English speakers, gyebale may be an especially useful term for acknowledging the collective contributions of as-yet-underappreciated members of *anthropos*, as well as extending appreciation for the underacknowledged but nevertheless vital work of a suite of other-than-human beings—microbes, fungi, insects, and many others—that continue to do work for us all.[11]

NOTES

1. Many thanks to David Schoenbrun for bringing this to my attention and to Olamide O Bisi-Amosun for explaining the use of this phrase in Yoruba for me.
2. Following Erving Goffman and Judith Irvine, some African linguists refer to these as passing greetings; see Femi Akindele, "A Sociolinguistic Analysis of Yoruba Greetings," *African Languages and Cultures* 3, no. 1 (1990): 1–14. Others instead consider these to be exhortatory greetings; for instance, Onuigbo G. Nwoye situates greetings "whose purpose or function is to admonish, or urge to greater or better performance." Nwoye, "An Ethnographic Analysis of Igbo Greetings,"

African Languages and Cultures 6, no. 1 (1993): 40. Kofi Agyekum explains that as an activity, greetings are meant to "encourage, praise and boost the morale of the addressee to continue to work hard." Agyekum, "The Pragmatics of Akan Greetings," *Discourse Studies* 10, no. 4 (2008): 506.

3. Euphrase Kezilahabi, "A Phenomenological Interpretation of Kerebe Greetings," *Journal of African Cultural Studies* 14, no. 2 (2001): 189.

4. Martin Walsh, "Nyakyusa Greetings," *Cambridge Anthropology* 7, no. 3 (1982): 32.

5. Onuigbo G. Nwoye, "An Ethnographic Analysis of Igbo Greetings," *African Languages and Cultures* 6, no. 1 (1993): 40.

6. Bernard Sabiiti, *Uglish: A Dictionary of Ugandan English* (Arlington, Mass.: J. C. Mugunga, 2014), 40.

7. Bebwa Isingoma, "Lexical Borrowings and Calques in Ugandan English," *Ugandan English: Its Sociolinguistics, Structure and Uses in a Globalising Post-protectorate*, ed. Christiane Meierkord, Bebwa Isingoma, and Saudah Namyalo (Amsterdam: John Benjamins, 2016), 167.

8. Agyekum, "Pragmatics of Akan Greetings," 506; Akindele, "Sociolinguistic Analysis of Yoruba Greetings," 11.

9. *Gyebale, gyebaleko,* and subsequent attestations of this term used in other African languages are most often used as an expression of appreciation and mutual respect in the sense described here. They can and sometimes are used by familiars in a more sarcastic register to reflect one's disappointment with something promised but not completed, or to guilt one clearly enjoying an undeserved rest into getting going on something that needs doing.

10. In Luganda: "Bwe mubanga mulya enswa: temuzitendanga kuwooma, nga temunnasaasira nkuyege ezaabumba ettaka."

11. Although microorganisms, fungi, and insects are too often described as pests and evoke feelings of disgust among English speakers, we literally would not be alive without them. By way of a few examples: our bodies contain more microbial cells than human ones, fungi gave us penicillin, and if it weren't for bees, we wouldn't have honey.

Pronunciation: heh-yi-yeh (hɛjɪjə)

Part of Speech: Noun

Provenance: Speculative fiction (Ursula K. Le Guin's 1985 novel *Always Coming Home*)

Example: There are innumerable expressions of heyiya. What they all share is a commitment to reshaping desire, forming ecological connections, and disconnecting from ecocidal forms of collective life.

Ursula K. Le Guin, the beloved author of novels, short stories, children's literature, and poetry, died in January 2018. After her passing, Le Guin's acceptance speech at the 2014 National Book Awards circulated widely, and her words from one moment in particular stood out: "We live in capitalism, its power seems inescapable—but then, so did the divine right of kings. Any human power can be resisted and changed by human beings."[1] Le Guin spent a lifetime imagining various forms of resistance. Fans and literary critics have devoted themselves to celebrated works like *The Left Hand of Darkness* (1969) and *The Dispossessed* (1974), but less attention has been given to what might be her greatest feat of the imagination. In *Always Coming Home* (1985), Le Guin creates a future world of unmatched breadth and artful precision. One can sense that something new is coming into being—something radically dissimilar to the global capitalist present. That emergent world takes shape through the practice of heyiya.

The Kesh are a people learning to inhabit California's Napa Valley in the distant future. Their Napa Valley is connected to our own: toxins, air and water pollution, and climate change from our present continue to impact their lives. The Kesh have human neighbors called the Dayao (also known as the Condor people), uncomfortably familiar figures that represent the present moment's ideological residuals. The Dayao exact violence upon humans they consider unlike themselves and against nonhumans whom they view as merely useful to their civilization. They attempt to accumulate scarce remaining resources to form a militant, imperial force. The Kesh respond to the ideological strain represented by the Dayao, along with the toxic and climatic residuals of our time, by living lives guided by heyiya.

Heyiya signifies a set of practices whose end is a vibrant ecology that flourishes in the Valley. Heyiya for the Kesh means "sacred, holy, or important thing, place, time, or event; connection; spiral, gyre, or helix; center; change."[2] This brief introduction to heyiya will maintain that it is a collective labor of imagining ourselves in and among the humans and nonhumans of this allegorical valley. Reading about how the Kesh practice heyiya provokes us to consider how our shared labor could be put to use in imagining a future desirable enough to begin letting go of our attachments to present systems of power. The aim of engaging in heyiya is ultimately to make our home anew through the labor of refashioning our desires.[3]

Practicing heyiya for the Kesh is learning how to become attached to ecologically viable visions and detached from destructive ones. Heyiya is an effort to recenter collective life on the joys of deeply felt relationship to other human animals, nonhuman animals, and living and nonliving material processes, with an openness to becoming attuned to the ecohistorical forces beyond their limited perception. It is a worldview that understands connection as intertwined with disconnection from the lures of commodification, colonialism, racialization, anthro-

pocentrism, and ecocide. Heyiya toggles between connection and disconnection, altering the form of Kesh material relations. This entry
will explicate how *Always Coming Home* models the practice of heyiya
in a manner that readers can emulate. I illustrate how the Kesh use
metaphors of the double spiral and the house to reimagine and rebuild
their relationships to their locality and the planet. The Kesh view the
dance and the hinge as metaphors of the libidinal energies needed to
reshape our relations of connection and disconnection to visions of
home. Finally, I turn to the mythical Coyote figure that teaches the Kesh
how to engage safely and creatively with the terror of ecocidal violence.

Always Coming Home reads more like a fragmented ethnographic
report than a novel.[4] A unique work in Le Guin's oeuvre, the novel for-
mally models heyiya, as the reader's pleasure comes from assembling
the various Kesh cultural practices that attune them to their neighboring humans, animals, plants, landscapes, and waterways. Sections of
the text are devoted to the Kesh's relations of kinship; to architecture and
the built environment, including maps of the Valley; to artistic sketches
of the guinea pig–like "himpi"; and to Kesh stories about significant animals and creeks. Still others include life stories and various characters'
poetry. *Always Coming Home* is not clear about how its diverse sections
fit together.[5] Readers' efforts to weave these genres, forms, and images
slowly reveal how the Kesh have figuratively and literally incorporated
their bioregion. The Valley starts to come into focus. Making sense of
the novel itself demonstrates that satisfaction can come from collective
efforts to grasp the various lifeways and ecological scales all around us.[6]

The fruits of laboring to imagine the Valley and its inhabitants are
valuable beyond our desire to understand a multiscalar world and its
complexities. The work of cocreating the Valley produces an affective
sense of possibility: the future world of the Kesh may be fictional, but
heyiya is real. Readers can begin to sense an intimate connection to life
in the Valley opening up a vision of a future that is not foreclosed by

current ecocical violence. Heyiya, then, is a collective, material process we can all engage in to find pleasure in ecosocial connection as well as the work of reversing overwhelming feelings of hopelessness and despair. Our world of capitalism and catastrophe, slow violence and dispossession, is not all that can be imagined and created.[7]

Readers of *Always Coming Home* slowly learn how heyiya becomes materialized in the Valley. The Kesh give shape to heyiya in a double spiral form that signifies its dialectical nature:

> It [*heyimas*] is formed of the elements *heya, heyiya*—the connotations of which include sacredness, hinge, connection, spiral, center, praise, and change—and *ma*, house. The heyiya-if, two spirals centered upon the same (empty) space, was the material or visual representation of the idea of heyiya. Varied and elaborated in *countless ways*, the heyiya-if was a choreographic and gestural element in dance, and the shape of the stage and the movement of the staging in drama were based upon it; it was an organisational device in town planning, in graphic and sculptural forms, in decoration, and in the design of musical instruments; it served as a subject of meditation and as an inexhaustible metaphor.[8]

The guiding source of metaphor, heyiya, is what connects the Kesh to each other and to other life in the Valley, but its manifestations take on "countless" and diverse expressions. This allows for unity and difference to coexist, along with opportunities for change. The passage above indicates that the Kesh understand heyiya as a sacred connection that comes into being when it takes on a material form in architecture, dancing, drama, poetry, and music. The Kesh recognize heyiya as a concept that should inform all expression, however artful or utilitarian. We could consider heyiya as a challenge to incorporate nonhumans into our creative, cultural, and political projects. Heyiya always points to relationships between ideas and material realities, symbolized by its double spiral: ideas, fantasies, and desires transformed into material realities, and vice versa.

The Kesh's double-spiral living arrangements imagine the construction of a more expansive vision of home. Le Guin writes: "In a Valley town everybody had two houses: the house you lived in, your dwelling-place, in the Left Arm of the double-spiral-shaped town; and in the Right Arm, your House, the heyimas. In the household, you lived with your kinfolk by blood or by marriage; in the heyimas you met with your greater and permanent family."[9] The left spiral represents human kin and is symbolically associated with everyday life: local animals and plants, geographic directions in the Valley, and soil and stones. The right spiral is where they build the *heyimas*, which represents the climate and the flows of the atmosphere, along with deep history and distant ecological impacts. The two spirals suggest that the Kesh understand these scales, the local and nonlocal, to be uneven but interconnected, requiring mediation in order to come into cultural focus.[10] We too might consider the urgency of cultivating knowledge and feelings about various ecological scales into cultural forms. What collective opportunities might emerge by creating community around the pleasures of ecological intimacies?

The Kesh have to practice dealing with scalar asymmetries and historical differences.[11] By doing so, the other beings of the Valley come into consciousness as normative members of the community. The Kesh's housing perpetually reminds them of their expansive kinship in the Valley. Nonhuman animals come to matter by bringing them "into the house," or by resignifying their relationships through naming and the production of space.[12] The Kesh's built environment situates them as a small part of the planet's material processes, but with the potential to, like us, radically reshape the geological, hydrological, and climatological systems. Today, built environments are tethered to a capitalism that reproduces racialized geographies, uneven development and resource appropriation, and neocolonial arrangements of accumulation and dispossession.[13] Heyiya, on the other hand, is a collective imaginative process that does not permit us to believe that contemporary systems of power are the only shape that history can take. Ecotopian processes like heyiya remind us that our labor can be put to other uses, our desires can build other lifeworlds, and our collective efforts can make that reality as rich and textured as the lives of the Kesh.

The Kesh create communal pleasures that help them to incorporate various scales of human and nonhuman difference. Dancing, for the Kesh, represents the corporeal and psychological pleasures of embodying heyiya. Each dance takes place in the same double spiral formation and meets at what they call the hinge. The hinge, like the hinge on a door, connects the internal and external by acknowledging differences between humans and animals, plants, and climatic-geological scales, but it also suggests they are bound together. A hinge allows for movement to avoid the ossification of desire. The Kesh view the Dayao's desires as fixated on ecosocial ways that are destructive. For the Dayao, only one leader, a single process, or a homogenous view of linear time is meaningful. From the Kesh perspective, this focus creates simplistic expressions of diversity and creativity; it also inadvertently produces a worldview that reinforces the Dayao sense of exceptionalism. The Dayao understand their place in the world as unique, chosen, and therefore worthy of the world's resources and energies. The Kesh call this kind of ideology and praxis living "outside the world." It is the same name they give to our present moment.

Unlike the Dayao, the Kesh have structured reversals into their cultural life to prevent the emergence of destructive desires. They believe that periodic shifts in cultural norms prevent anthropocentric and ethnocentric behaviors from taking root while simultaneously promoting the pleasures of connection and disconnection as equally necessary:

At the World Dance people get married; that's a wakwa
[dance, spring, or water source] of sorting out things, getting
things right and flowing on the two sides of the world; that's
a wakwa of lasting and staying. The Moon Dance doesn't do
anything like that. It goes the other way. It goes out and apart,
undoing, separating. You know the heyiya-if comes in to the
center and at the same time it's going out from the center.
A hinge connects and it holds apart. So under the Moon
there are no marriages. No households.[14]

Kesh dances are a shared celebration of the connections between
human animals and the various ecological and cosmological scales.
Dances mark large-scale movements of shifting seasons and celestial
objects, along with the everyday scales of the community's nonhuman members such as bears, deer, pumas, coyotes, and hawks. How
would our perceptions of the climate, for instance, become altered if
we danced its many forces, like the Kesh, feeling them in our bodies?
They dance to make these scalar forces visceral, memorable, familiar,
and exciting. The Kesh also relish changing direction at the hinge. In
the passage above, the movement to the center of the heyiya-if (arriving
at the hinge) is about a change in motion from connection and duration
to opening up for separation. Hinge acts acknowledge that we must be
reflexive and open to reversals if we are to find collective pleasure in
reversing the calamities of our present: climate change, mass extinction, racial and ethnic conflict, and refugee crises arising from military
excursions and ecological distress. We can become more open to letting
go of, disconnecting from, or reversing desires that contribute to this
suffering. Knowing what to mourn—what relationships and structures
to let go of—has never been more urgent.

The center of heyiya is where the Kesh hinge, but as the first passage made evident, they also understand this focal point of the heyiya
as perpetually empty. The clearest explanation for this apparent paradox

is that the empty center represents what remains unresolved in history, a contradiction or antagonism that shapes Kesh life. The Kesh call this living outside the world. They associate the ecocidal behaviors of living outside the world with the past (our present) and with the Dayao. The Kesh gain their cultural identity from this contrast, which has the potential to turn violent when an "other" is found to blame. They must therefore sublimate the tendency to turn to warfare if they are to remain true to the ethos of heyiya. Kesh people dance to find better use for these energies, but they also tell stories about the outside that both diffuse the urge to harm and reorient unresolved conflict toward creative ends.

The Kesh continually fill the void at the center of heyiya with stories that create new life out of mass death and destruction. The last two elements of heyiya are, not coincidentally, "center" and "change." The Kesh reference this empty center of heyiya through stories of recreating the world by consuming and transforming the "outside." These tales include the birth of the Valley, which emanated from the mythical Coyote figure's shit after she had devoured the poison of the reader's present. One set of stories lets the Kesh spiral back to the empty center to engulf the toxicity of the "outside" and thus to keep working toward building up the "inside." Coyote's myths do not lend themselves to self-aggrandizement—in the manner of the Dayao—as if the Kesh were uniquely important. Rather, they reverse this tendency, narrating the Valley and all of its members, human and nonhuman, as born of Coyote's excrement. These stories depict the "outside" as a past period, and therefore empathetic space opens up for the Dayao, who have received this historical inheritance just like the Kesh. Both groups were born of the past's shit. At other times, representing the second spiral of heyiya, they tell horrific stories like "A Hole in the Air," in which a man enters the "outside of the world" through a hole in the air and continues to die over and over from grief and poison. Both types of origin stories allow the Kesh to see their present as a material and ideological remnant. The

Kesh prefer to place blame on historical processes, not people. We too can work against systems of power that are destructive without forgetting that its perpetrators have learned how to desire from the "outside," and it is the systems that must change. The Kesh ultimately realize through their unresolved conflicts that the work of truly living in the Valley is not yet complete.

The Kesh therefore do not merely live in the Valley; they desire to live in and among it more fully. Heyiya is the enjoyment of seeking the Valley together, knowing that in true heyiya fashion—with its perpetually empty center—the desire to fully inhabit the Valley is perpetually on the horizon. We will need to dance again and craft new tales. As the narrator of *Always Coming Home* tells us, "We'll go on, and you'll hear the quail calling on the mountain by the springs of the river, and looking back you'll see the river running downward through the wild hills behind, below, and you'll say, 'Isn't that it, the Valley?' And all I will be able to say is, 'Drink this water of the spring, rest here awhile, we have a long way yet to go, and I can't go without you.'"[15] *Always Coming Home* suggests that we too can build a home in the Valley. This great work is one we can enjoy together if we are willing to see it as a work of constant imperfection, only to spiral back and try again. Such efforts can also detach us from normative expectations that scientists, politicians, or geoengineering will somehow save us. Like the Kesh, it takes the collective power of the group, not a singular hero. The practice of reading ecotopian fiction such as *Always Coming Home* can inspire a sense that the future is not foreclosed to our political, economic, and cultural interventions. The world becomes ours to remake when we join the dance, practice hinge acts, build the house, and tell Coyote stories. Take a moment to look around, because "we have a long way yet to go, and I can't go without you."

Fotminne
ANOTHER PATH

NOTES

1. Marissa Martinelli, "Remember the Late Ursula K. Le Guin by Rewatching Her Remarkable Speech at the 2014 National Book Awards," Slate, January 23, 2018, https://slate.com/.
2. Ursula K. Le Guin, *Always Coming Home* (1985; reprint, Berkeley: University of California Press, 2001), 515.
3. The "education of desire" has long been considered by critical theorists to be fundamental to utopia literature. See, for instance, Frederic Jameson, *Archaeologies of the Future: The Desire Called Utopia and Other Science Fictions* (London: Verso, 2005); Ruth Levitas, *The Concept of Utopia* (Syracuse, N.Y.: Syracuse University Press, 1999); and Christine Nadir, "Utopian Studies, Environmental Literature, and the Legacy of an Idea: Educating Desire in Miguel Abensour and Ursula K. Le Guin," *Utopian Studies* 21, no. 1 (2010): 24–56.
4. Le Guin acknowledges that her anthropologist parents influenced her formal choices in this novel. She has been asked if Native American cultures influenced the creation of the Kesh. She found that Native American oral literature helped her to capture aspects of the Napa Valley. She insists, however, "I certainly didn't want to put a bunch of made up Indians into a Napa Valley of the future." There is certainly room for concern about resemblances that could be read as what Shepard Krech III calls the ecological Indian myth. Simultaneously, the novel exhibits anarchist, Marxist, Taoist, and ecofeminist influences and Le Guin's own express concerns about "stealing or exploiting [Native American literature], because we've done enough of that to Native American writing." See Jonathan White, *Talking on the Water: Conversations about Nature and Creativity* (San Antonio, Tex.: Trinity University Press, 2016), 117; and Krech, *The Ecological Indian: Myth and History* (New York: Norton, 2000).
5. Le Guin does signal to readers, however, through her fictional alter ego, Pandora. It is Pandora who collects and curates the novel. She writes the following at the end of a short note entitled, "Pandora Worries about What She Is Doing: The Pattern": "Even if the bowl is broken (and the bowl is broken), from the clay and the making and the firing and the pattern, even if the pattern is incomplete (and the pattern is incomplete), let the mind draw its energy. Let the heart complete the pattern." Le Guin, *Always Coming Home*, 53.
6. Donna Haraway's work on "string figures" and "tentacular thinking" utilizes speculative fiction in a critical and creative approach to various scales that has some resonance with heyiya. See Donna Haraway, *Staying with the Trouble: Making Kin in the Chthulucene* (Durham, N.C.: Duke University Press, 2017).
7. On the concept of "slow violence," see Rob Nixon, *Slow Violence and the Environmentalism of the Poor* (Cambridge, Mass.: Harvard University Press, 2011). On the

valences of dispossession, see David Harvey, "The 'New' Imperialism: Accumulation by Dispossession," *Socialist Register* 40, no. 1 (2004): 63–87.

8. Le Guin, *Always Coming Home*, 45; emphasis added.

9. Le Guin, *Always Coming Home*, 48.

10. On the necessary mediation of ecological phenomenon, see Timothy Morton, *Ecology without Nature* (Cambridge, Mass.: Harvard University Press, 2009), and Jesse Oak Taylor, "The Novel as Climate Model: Realism and the Greenhouse Effect in *Bleak House*," *Novel* 46, no. 1 (2013): 1–25.

11. For a discussion of the contemporary moment's derangement of scale, see Timothy Clark, "Scale," in *Telemorphosis: Theory in the Era of Climate Change, Vol. 1*, ed. Tom Cohen (Ann Arbor: Open Humanities Press, 2012), http://www.open humanitiespress.org/books/titles/telemorphosis/.

12. On foundational ideas about the production of space, see Henri Lefebvre, *The Production of Space*, trans. Donald Nicholson-Smith (New York: Wiley-Blackwell, 1992).

13. For recent scholarship on racial geographies, see María Josefina Saldaña-Portillo, *Indian Given: Racial Geographies across Mexico and the United States* (Durham, N.C.: Duke University Press, 2016), and for the central role of capitalist appropriation, see Jason Moore, *Capitalism in the Web of Life: Ecology and the Accumulation of Capital* (New York: Verso, 2015). For an introduction to uneven development, see Neil Smith, *Uneven Development: Nature, Capital, and the Production of Space*, 3rd ed. (Athens: University of Georgia Press, 2008).

14. Le Guin, *Always Coming Home*, 242.

15. Le Guin, *Always Coming Home*, 339.

Hyperempathy

Rebecca Evans

Pronunciation: hi-per-em-path-ee (hī-pər-*em*-päth-ē)

Part of Speech: Noun

Provenance: Speculative fiction (Octavia E. Butler's Parable series)

Example: In that tragic era of extinction her hyperempathy was sometimes a burden, but it motivated her campaign of direct action, which—some historians argue—precipitated the Great Transition to the world we enjoy today.

The Anthropocene identifies external transformation—of oceanic chemistry, of coastlines, of habitats—but it also suggests the internal of human experience. How do we think about our relationships to global environments and resources, and how must we learn to think differently? What scenarios and possibilities should we steel ourselves for, and what can we dare to imagine? What habits of feeling might help "us"—understood across scales, as local, regional, global, and even multispecies communities—survive? The inherent communality of human life is at the heart of each of these questions. This entry explores empathy as a possible tool for inspiring just action in the face of climate change.

I'll begin with a story of empathy in action. In the second chapter of Octavia E. Butler's dystopian 1993 novel *Parable of the Sower*, the teenage narrator, Lauren Olamina, recounts her morning adventure: joining a small group of family and friends as they bicycle beyond the walls of her gated Southern California community, Robledo. The group is heading to a nearby church where her father was a pastor before it became too dangerous to travel beyond the gated community; they're taking the risk of leaving on this particular day because Olamina and the other children are scheduled to be baptized. As she bicycles, Olamina tries her best not to look too closely at the carnage evident in the ravaged landscape beyond Robledo. At first, this seems like fairly normal behavior for a teenage girl as she struggles to avoid succumbing to fear: "One of them was headless," Olamina writes; "I caught myself looking around for the head. After that, I tired [sic] not to look around at all" (9).[1] This resolve, however, doesn't hold. "As I rode," Olamina continues, "I tried not to look around at them, but I couldn't help seeing—collecting—some of their general misery" (11). With the strangeness of phrasing—how precisely does one collect misery?—readers begin to wonder about the reasons behind Olamina's practiced avoidance of pain. Olamina writes: "I can take a lot of pain without falling apart. I've had to learn to do that. But it was hard, today, to keep peddling and keep up with the others when just about everyone I saw made me feel worse and worse" (11). Finally, Olamina reveals that this is not a dramatic account of the general workings of empathy but an unusual psychological disability: hyperempathy, or "what the doctors call an 'organic delusional syndrome'" (12). Because Olamina's mother used a prescription drug whose side effects were unknown at the time, Olamina was born a hyperempath, meaning she physically experiences any pain or pleasure she observes in others. In typical teenage fashion, Olamina makes light of her condition: "Big shit. It hurts, that's all I know," she says. "Anyway, my neurotransmitters are scrambled and they're going to stay scrambled" (12). But the toll it takes on her shines fiercely through her adolescent insouciance: the bicycle rides, she confesses to her diary, "were hell . . . the worst things I've ever felt—shadows and ghosts, twists and jabs of unexpected pain" (13).

In Octavia Butler's world, life with hyperempathy is a state of constant low-level shame and anxiety. Sharers, as those with hyperempathy are called, are vulnerable, experiencing not only the pain they inflict but also the pain they merely witness; pleasure is a less frequently discussed component of the condition. Even fake pain sets Olamina off. When she was a child, her younger brother covered himself in red ink to trick her into bleeding in empathic response (*Sower* 11). Those with hyperempathy live their lives on a spectrum between overprotection and overexposure. As Olamina's brother tells her after a year spent beyond the borders of Robledo, "Out there, outside, you wouldn't last a day. That hyperempathy shit of yours would bring you down even if nobody touched you" (*Sower* 110). Even beyond these pragmatic concerns, the fact that sharing is produced by maternal drug use means that sharers are stigmatized—even though, as Olamina tells us in the second Parable book, the drug that caused her condition was a common prescription medication akin to Adderall (*Talents* 18).[2] A combination of vulnerability and shame weighs on sharers. "I've never told anyone," Olamina reflects, long after she has left Robledo, as she contemplates revealing her condition to her fellow travelers: "Sharing is a weakness, a shameful secret. A person who knows what I am can hurt me, betray me, disable me with little effort" (*Sower* 178).

Her status as a sharer clearly shapes Olamina's character, but hyperempathy seems at first to be incidental to the larger themes— particularly the ecological ones—of Butler's Parable books. *Parable of the Sower* and *Parable of the Talents*, two of the most famous and beloved environmental science fiction books of the last few decades, are not ultimately stories about hyperempathy. Instead, most of their energy is dedicated to tracing the contours of the novels' dystopian near-future America—marked by climate change, resource shortages, economic collapse, and the rise of violent fundamentalism—and to following Olamina as she founds a new religion, Earthseed, which aims to form more adaptable and sustainable communities on Earth while laying the groundwork for those communities to travel to other planets and "take root among the stars" (*Sower* 77). The bulk of the two novels focuses on the evolution of Earthseed from a vague concept with which the younger Olamina is preoccupied to a full-fledged religion that secures widespread prestige, masses of followers, and, as *Parable of the Talents* ends, funding for an interstellar colonizing mission. The path is hardly smooth. In the first novel, Olamina is forced to flee Robledo when it is razed, and she undergoes suffering, danger, and loss on the road north from Los Angeles before finally founding the first Earthseed enclave, Acorn, in Northern California. In the second novel, Acorn, which has developed into a thriving Earthseed community, is invaded by recently empowered fascist Christian fundamentalists. Most of its inhabitants are killed. Others remain enslaved until Olamina spearheads a revolt and eventually rebuilds Earthseed. Hyperempathy guides Olamina's choices and reactions in moments of immediate danger, but Butler doesn't foreground sharing as central to Earthseed or to the dystopian conditions in which Olamina lives.

Yet hyperempathy is a key part of both Olamina's and Butler's ecological projects. One might, in other words, see sharers not as a tangential departure from Butler's (and her protagonist's) efforts to imagine a way out of violence, environmental degradation, and injustice. Instead, one might view hyperempathy as the motivation for and core of Olamina's vision of climate justice. The intimate relationship between hyperempathy and Earthseed explains why hyperempathy belongs in Butler's critical dystopia, and why only a sharer like Olamina could build a path from dystopia to utopian possibility.[3] Olamina's resistance to environmental and social injustice, and her vision for egalitarian and ecotopian futurity, are each rooted in her status as a hyperempath. Moreover, hyperempathy marks both the productive and the deactivating aspects of consuming environmental media in the Anthropocene. What are the possible relationships between empathy and ecological justice? Does being empathic necessarily make one averse to causing pain, or, problematically, can it make one merely averse to witnessing it? When do (hyper)empaths tend toward action, as opposed to stasis? Olamina functions as a useful narrative laboratory in which to investigate these questions—questions that are not only theoretically engaging but are in fact critical to the way we produce, consume, and navigate media in the Anthropocene. In a moment of resurgent interest in the Parable books, including a new operatic adaptation by Toshi Reagan that premiered in 2017, hyperempathy seems poised to play a significant role in contemporary conversations about resistance, community, justice, and sustainability.

Twice in the series—once in each book—a narrator posits a psychological link between Olamina's hyperempathy and her unnervingly unwavering commitment to Earthseed. Early in *Parable of the Sower*, Olamina herself questions why she "can't do what the others have done—ignore the obvious. Live a normal life," why Earthseed "won't let [her] alone, won't let [her] forget it, won't let [her] go." "Maybe. . . . Maybe it's like my sharing," she thinks: "One more weirdness; one more crazy, deep-rooted delusion that I'm stuck with. I am stuck with it" (26). Strikingly, this is the only time Olamina acknowledges what may seem like an evident connection between her defining characteristics. In this formulation, utopianism and hyperempathy are both delusional and inevitable: you have them or you don't. Olamina marks herself here, as she does frequently in the Parable books, as singular, extraordinary. She may recognize her qualities as off-putting and rare, but she doesn't seem able to consider the common seed from which these qualities grow.

In *Parable of the Talents*, however, Olamina's estranged daughter, Larkin, offers a less exceptionalist account of why Olamina so singlemindedly dedicated her life to Earthseed's success: "Well, my mother was a sharer, a little adult at 15, and a survivor of the destruction of her whole neighborhood at 18. Perhaps that was why she . . . needed to take charge, to bring her own brand of order to the chaos that she saw swallow so many of the people she loved" (103). Other characters tend to echo Larkin's perspective, articulating Earthseed and sharing not as isolated expressions of Olamina's idiosyncrasy but as causally linked: the experience of debilitating trauma produces the desire for a community that more equitably distributes (and more insistently minimizes) pain. As Olamina's partner, Bankole, says, "It might not be so bad a thing if most people had to endure all the pain they caused" (*Sower* 277–78). Though Olamina openly rejects Bankole's formulation as naive, his comment echoes an earlier reflection of hers, a record of her grief after her brother is found murdered:

> It's beyond me how one human being could do that to another. If hyperempathy syndrome were a more common complaint, people couldn't do such things. They could kill if they had to, and bear the pain of it or be destroyed by it. But if everyone could feel everyone else's pain, who would torture? Who would cause anyone unnecessary pain? I've never thought of my problem as something that might do some good before, but the way things are, I think it would help. I wish I could give it to people. Failing that, I wish I could find other people who have it, and live among them. A biological conscience is better than no conscience at all. (*Sower* 115)

This formulation of the sharer as a person with a "biological conscience" is striking: it naturalizes politics and ethics, relying on affect and involuntary physiological response to accomplish what culture and morality do not. Hyperempathy offers both ecologically oriented consciousness and ecologically rooted conscience.

Olamina's experience with hyperempathy thus suggests that empathy itself—specifically, the empathy that arises from watching someone else suffer[4]—can produce both more socially just and more ecologically sustainable communities. Olamina consistently notes that hyperempaths naturally resist individual violence: "Sharing pain . . . makes us very slow to cause pain to other people. We hate pain more than most people do" (*Talents* 37). Yet hyperempaths would presumably suffer equally from witnessing those forms of pain that are disproportionately inflicted by structural, indirect, or slow violence—forms of pain that, as the Parable books insistently record, climate change is bound to produce.[5] Sharers don't require malicious infliction in order to feel pain as their own; they feel all kinds of pain, not just pain inflicted by spectacular acts of violence. This suggests that the only society in which a sharer could be safe and secure is a society in which all forms of pain are scrupulously avoided, regardless of social status or even species. Hyperempathy thus literalizes philosopher John Rawls's "veil of ignorance." Rawls suggests that citizens can make truly moral decisions only if they lack previous knowledge of their social position.[6] In other words, people cannot make a moral decision unless they would be willing to accept the consequences of that decision from any position—as the most or as the least privileged member of society. Like the citizen behind the veil of ignorance, the sharer can't rely on a position of privilege: if suffering exists to be witnessed, she will experience it. She therefore must strive to create a world in which suffering is avoided whenever possible.

Hyperempathy also materializes a solution to the problematic politics of sympathy in environmental media. In some ways, the sharer performs what Adam Smith describes in *The Theory of Moral Sentiments* (1759) as the workings of sympathy: she experiences that which she witnesses others experiencing.[7] Yet Smith's formulation of sympathy opens up troubling gaps: not only are we are more likely to sympathize with those we already recognize as akin to us, but also sympathy (feeling-for) is a more mediated and malleable emotion than empathy (feeling-with). If apathy operates in the third person ("somebody else is suffering"), sympathy operates in the uneasy speculative first person ("what if I were the one who suffered?"). Empathy, by contrast, operates in the first person ("I am suffering"). As South African novelist J. M. Coetzee puts it in his metafictional novel *The Lives of Animals* (1999), sympathy "allows us to share at times the being of another," but ultimately it "has everything to do with the subject and little to do with the object."[8] Hyperempathy closes this gap: for Olamina, there is no such thing as critical distance from perceived suffering.

Sharers embody not sympathy but radical empathy, which prompts moral action and avoids a reliance on a distant, speculative, and often pitying relation to those who suffer. Hyperempathy thus exemplifies an emotional orientation that may be necessary to catalyze action in the face of environmental and climate injustice. Olamina's status as a hyperempath is not incidental to her ecological politics; instead, her unyielding goal of a just, sustainable, and adaptable society is a response to her hyperempathy. The sharer thus stands in for the ideal audience of environmental media: a reader or viewer who, rather than responding with pity or apathy, is moved to act as though—or indeed because—the suffering she witnesses is understood and experienced as her own. Within the logic of Butler's science fiction, this state must be conferred inexorably from without, a neurochemical deus ex machina, and Butler was certainly cynical about humankind's capacity to craft better worlds without such imperatives.[9] Yet read through the lens of what literary theorist Fredric Jameson, after Marxist philosopher Ernst Bloch, calls the utopian impulse (rather than the blueprint for utopian future often associated with the genre of utopia), the novels give us not a solution but instead a shared desire to feel better—to feel more deeply, to be guided by gut feelings with less cynical reserve or self-interest.[10] Hyperempathy represents a transformative form of feeling as we receive others' pain, a way to live in the unevenly distributed violence of the Anthropocene without becoming detached or complacent, a necessary habit of acting to prevent the pain of vulnerable others as surely and instinctively as you would act to prevent your own pain.[11]

At the same time, however, Butler's Parable books also illuminate the problems posed by a sharer, the aspects of hyperempathy that might in fact detract from environmental and climate justice in the

Anthropocene. As Olamina writes, "It is incomprehensible to me that some people think of sharing as an ability or a power—as something desirable" (*Talents* 37). Butler's sharers face a number of logistical difficulties. They are unreliable in dangerous situations where another's pain might disable them, and knowledge of their vulnerability makes them easy targets for exploiters and bullies, making even self-defense a difficult task (*Sower* 278).[12] Further, Olamina's hyperempathy also runs the risk of making her avoidant, escapist, and calculating. Indeed, Olamina's controversial commitment to leaving Earth behind and starting anew on other planets might be seen as an extension of her hyperempathy. After all, might sharers not prefer the immediate fix of retreating from visible pain to the more arduous task of working to craft a less painful social order? Butler's novels raise questions about the assumption that painful empathy should play a central role in environmental representation: does bombardment with abjection and suffering make somebody more likely to take action, or more likely to look away?[13] Are hyperempaths likely to succumb to apathy, to the NIMBYistic desire to shield (only) one's immediate community from the effects of environmental injustice?

Butler's Parable series thus asks us to understand empathy's role in crafting environmental outcomes. Olamina demonstrates how (hyper) empathy can function both as an asset and as a hindrance to the pursuit of environmental justice. The sharer's experience of the disproportionate violence of climate change is urgent and immediate; it is also capable of producing utopian commitments. Yet that experience can also produce apathy, avoidance, vulnerability, overstimulation, and uncomfortable divisions along gender lines. Hyperempathy requires us to consider how the cultural deployment of pain can prompt a counterintuitive retreat from environmental engagement even as it holds out the possibility of transformational structures of feeling.[14]

Shikata ga nai

ANOTHER PATH

NOTES

1. References to this text will be made using parenthetical citation: Octavia E. Butler, *Parable of the Sower* (1993; reprint, New York: Grand Central, 2007).

2. References to this text will be made using parenthetical citation: Butler, *Parable of the Talents* (New York: Seven Stories, 1998).

3. In *Scraps of the Untainted Sky: Science Fiction, Utopia, Dystopia* (Boulder, Colo.: Westview Press, 2000), science fiction studies scholar Tom Moylan defines "critical dystopia" as a hybrid genre that depicts a dystopian world but contains the embedded promise of utopian possibility that could be achieved. This characterizes Butler's Parable books: while they depict a dystopian California, Olamina's eventually successful work to create the Earthseed movement always embeds the hope and eventually promises the fruition of a just and sustainable community.

4. Olamina rarely reflects on her experience of other people's pleasure. This gap has interesting resonance for contemporary environmental discourse, which often focuses on pain, suffering, and disaster, actual or projected. In the context of the Anthropocene's human and other-than-human suffering, an emphasis on pain makes sense. As Butler's novels seem to suggest, pain is what produces the strongest experience of empathy and the most urgent impetus to action.

5. The term "structural violence" was proposed by Johan Galtung in the late 1960s to describe any disproportionate or unnecessary experience of death, pain, or deprival; the concept was inspired by Stokely Carmichael and Charles V. Hamilton's concept of institutional racism. These forms of violence also characterize environmental racism; meanwhile, as Rob Nixon has argued, climate change enacts "slow violence" against the global poor. See Galtung, "Violence, Peace, and Peace Research," *Journal of Peace Research* 6, no. 3 (1969): 167–91; Carmichael and Hamilton, *Black Power: The Politics of Liberation in America* (New York: Random House, 1967); and Nixon, *Slow Violence and the Environmentalism of the Poor* (Cambridge, Mass.: Harvard University Press, 2011).

6. The veil of ignorance requires citizens to design society from the original position (without knowing where they will land within the social order). As Rawls clarifies, the original position is not "an actual historical state of affairs" but rather "a purely hypothetical situation." Rawls, *A Theory of Justice*, rev. ed. (Cambridge, Mass.: Harvard University Press, 1999), 11.

7. Smith understands sympathy as the "moral sentiment" that allows humans to experience an echo of what another must be feeling through the workings of the imagination. He describes it as intensely visual, "the emotion which we feel for the misery of others, when we either see it, or are made to conceive it in a very lively manner." Smith, *A Theory of Moral Sentiments* (1759; reprint, New York: Arlington House, 1969), 3. Similarly, Olamina only shares the things she sees; hearing

somebody screaming in pain makes her nervous but has no hyperempathic effect (*Sower* 132).

8. J. M. Coetzee, *The Lives of Animals* (Princeton, N.J.: Princeton University Press, 1999), 34–35.

9. As Gerry Canavan observes, the published Parable books and many incomplete drafts toward a third book from her archives in the Huntington Library evince both Butler's cynicism about social and environmental devastation and her hope that even at the brink of catastrophe humans might ultimately be capable of change. Canavan, *Octavia E. Butler* (Champaign: University of Illinois Press, 2016).

10. As Jameson writes, the utopian is "not the commitment to a specific machinery or blueprint, but rather the commitment to imagining possible Utopias as such, in their greatest variety of forms." Jameson, *Archaeologies of the Future: The Desire Called Utopia and Other Science Fictions* (New York: Verso Books, 2005), 217.

11. Environmental justice cultural studies scholars are keenly aware of the role that empathy must play when transmitting or receiving the suffering of others. As Michael Ziser and Julie Sze put it, culture has the "privileged ability to articulate differences in worldview, facilitate mutual understanding, and even trigger the empathy that lies at the heart of global environmental justice." Ziser and Sze, "Climate Change, Environmental Aesthetics, and Global Environmental Justice Cultural Studies," *Discourse* 29, no. 2–3 (2007): 386. See also Sze, "Environmental Justice Anthropocene Narratives: Sweet Art, Recognition, and Representation," *Resilience* 2, no. 2 (2015); and Nicole Seymour, *Strange Natures: Futurity, Empathy, and the Queer Ecological Imagination* (Champaign: University of Illinois Press, 2013).

12. In his childhood, Olamina's brother innocently comments "what good slaves sharers would make," and it turns out that slavers pay more for hyperempathic children (*Sower* 300, 305). In fact, as the fundamentalist coup enables more and more slave labor, slavers develop a collar that controls the wearer's pain, and Olamina notes the eerie similarity between this technology of violence and her own condition (*Talents* 80).

13. For more on representation and empathy from literary studies, see Martha Nussbaum, *Poetic Justice: The Literary Imagination and Public Life* (Boston: Beacon Press, 1995), and Suzanne Keen, *Empathy and the Novel* (Oxford: Oxford University Press, 2007). However, scholarship on the failure of dire predictions to prompt action on climate change suggests that depressing dystopian visions of climate futures may not be particularly effective in terms of producing environmental action. See Per Espen Stoknes on the deactivating "Great Grief" of climate change in *What We Think About When We Try Not to Think About Global Warming: Toward a New Psychology of Climate Action* (White River Junction, Vt.: Chelsea Green, 2015), 172; Adam Seth Levine and Reuben Kline on the distinction between public opinion

and political action in "When Does Self-Interest Motivate Political Engagement? The Case of Climate Change," *SSRN*, March 12, 2017, https://ssrn.com/abstract =2931842; and Matthew Feinberg and Robb Willer on the counterproductive effects of catastrophic climate discourse in "Apocalypse Soon? Dire Messages Reduce Belief in Global Warming by Contradicting Just-World Beliefs," *Psychological Science* 22 (2011): 34–38.

14. Raymond Williams, in *Marxism and Literature* (Oxford: Oxford University Press, 1978), introduced this term in the 1970s to describe the social nature of subjective experience.

Ildsjel

Karen O'Brien and Ann Kristin Schorre

Pronunciation: il-shail ('ilʃeːl)

Part of Speech: Noun, plural (ildsjeler)

Provenance: Norwegian

Example: The new community recycling project owes its existence to Shevek, a passionate ildsjel.

Climate change, biodiversity loss, ocean acidification, and deforestation are transforming socioecological systems, with the possibility of crossing thresholds or "planetary boundaries" that provide a safe operating space for humanity.[1] Given what is at stake, it is useful to step back and reflect on our ability to consciously produce a world where all species and ecosystems can thrive. Do we have the capacity to transform society rapidly, equitably, and sustainably? What role do individuals play in the process of social, political, and ecological transformations?

If our goal is to deliberately reduce the risks of climate disruptions and other threats to sustainability, we need to unleash humanity's potential for both individual and collective change. The word *agency* is widely used to describe a person's capacity to act independently and make free choices to influence outcomes. Without agents of change, there is little hope of creating societal transformations at the rate, scale, magnitude, and depth that scientists and policy makers consider necessary to avoid the most dangerous climate scenarios. Unfortunately, the words *agent* and *agency* fall flat; they do little to inspire action or to connect individuals to the communities of which they are a part. We need a livelier word that captures and activates the capacity of agents and agency to generate radical social change. The Norwegian word ildsjel does just that.

A BETTER WORD FOR A BETTER WORLD

Ildsjel is a noun that has been loosely translated into English as "enthusiast," or someone with a passionate commitment to a cause. Ildsjeler (the plural of ildsjel) work against the odds, striving for goals that they—and their communities—consider important, regardless of the economic, social, and structural obstacles. They are innovators who often take risks and do things that others consider impossible. More than mere dreamers, ildsjeler tend to be visionaries driven by possibilities and potentials for transformation. Of course, desirable visions tend to be normative, and they will vary between groups and situations, depending on the values and interests of the ildsjel. When it comes to transformations to sustainability, it is important to pay attention to the social context and espoused values of ildsjeler.

Let's consider the social context in which the term is currently used in Norway, a social democratic state with cultural values that emphasize solidarity and equity for all. Rooted in the trade union movement of the late nineteenth century and the struggle for universal suffrage, Norway's social democracy was built around a welfare state that aimed for full employment, considerable state control over the economy, and the redistribution of wealth. Norwegian culture emphasizes codetermination and egalitarianism; it recognizes that social democracy depends on a shared sense of community.[2]

In Norway, ildsjeler are celebrated for the positive role that they play in groups, organizations, and initiatives, especially through voluntary service. They are active in schools, sports, the arts, and festivals, as well as in community development projects, local history and museum projects, and innovative businesses.[3] Ildsjeler are those who do more than what is expected of them within a culture that values solidarity and where altruistic actions are appreciated.

In Norway, numerous books and articles have been written about ildsjeler, and awards are given annually to people with a passion and commitment to doing good. On the basis of interviews with ten ildsjeler in Norway, journalist Niels Christian Geelmuyden reports that they tend to live for meaning and purpose, contributing to something greater than themselves.[4] Examples of ildsjeler are plentiful, though many of them work behind the scenes and are seldom recognized for their roles outside of their communities. The term ildsjel applies not only to well-known activists such as Martin Luther King Jr. and Gandhi, but also to the man who organizes a movement to protect a local park, or the woman who, despite the obstacles, perseveres to design an inexpensive water purifier that protects millions of people from cholera.

Ildsjeler make change happen, often by inspiring others to join the process. In an interview with Geelmuyden, Ingebrigt Steen Jensen says: "The fire must benefit others. . . . An *ildsjel* must get the fire to spread. *Ildsjeleri* does not occur in isolation."[5] Ildsjeler stoke the fires in other souls; they create spaces that attract people and inspire participation and engagement. But what does this word bring to the table that is missing in the current lexicon of transformative change?

AGENTS OF CHANGE

The concept of agency refers to the capacity of individuals and groups to play a significant role in shaping outcomes that influence their futures. To have personal or social agency is to be capable of exerting power to achieve some end or goal. This includes the potential to exercise free will and to act against existing social structures and/or create new ones. Agency also implies some degree of self-awareness and reflexivity, including an ability to question existing norms and challenge the status quo. Robert Kegan and Lisa Laskow Lahey, experts in adult learning and education, describe the development of agency as a shift from a socialized mind-set to a self-authoring mind-set.[6] This especially involves challenging the social and cultural scripts as well as the assigned roles and identities that are perceived or presented as immutable or unquestionable.

The need for a critical, reflective, and nuanced approach to the concept of human agency is important, as individual and social agency have undeniably contributed to climate change and many other acute social and environmental problems. Throughout history, the agency of individuals and groups has been used to dominate, oppress, or dispossess others, often by expropriating land, exploiting resources, or contaminating air, water, or soil. Some individuals and groups bear a greater responsibility for the greenhouse gas emissions that contribute to climate change, and some have more power to influence or impede solutions. Indeed, some people may use their agency to protect vested interests in fossil fuel extraction and consumption, labeling those who challenge the status quo as dangerous and disruptive to energy security.

Scientists argue that humans now play such a significant role in shaping Earth system processes that we have entered a new geologic epoch, the Anthropocene.[7] Some point to population growth as a key driver of human impacts on the environment, while others emphasize population, affluence, and technology—the famous I = PAT equation.[8] Common to these theories is a recognition that humans are using resources and transforming our natural and social worlds at an unprecedented rate, magnitude, and scale.

To gain a better understanding of the collective behavior of humans, interdisciplinary teams of researchers have developed models that explore how simple rules can generate complex behaviors observed in society. Agent-based modeling is a method that simulates the actions and interactions of autonomous agents, whether they are individuals or collective organizations and groups.[9] In these models, human agents are presumed to be rational and acting in their own self-interest, yet capable of learning and adapting. This approach assumes that when we realize that an ecologically sustainable way of life is in our own self-interest, we will respond rationally and change behaviors, develop new technologies, and collaborate on local and national policies and international agreements.

Yet conscious decisions and intentional actions taken by individuals and groups are both influenced by and affect social norms, rules, regulations, institutions, and laws, as well as cultural beliefs and worldviews.[10] Solving environmental problems requires more than behavioral nudges. It requires stimulating collective agency to transform social structures and power relationships that have negative impacts on both people and the planet.

It is, however, difficult to act truly autonomously and deliberately step out of existing social norms and practices, particularly those that are embedded in culture. Indeed, many of those who are recognized as agents of change are always to some extent operating within a particular social discourse. An example is the climate activist who focuses on changing people's habits to reduce carbon emissions while failing to consider the underlying political, economic, social, and cultural conditions that maintain the logic of consumerism—what can be thought of as "the habits of capitalism."[11] Can we think outside the box when it comes to solutions to complex problems such as climate change, where the future is dependent on the current choices of a multitude of human actions? Do the concepts of agent and agency inspire people to transform?

In everyday language, the word *agent* is not particularly exciting. It has no soul, no energy, and a limited capacity to generate positive interventions. The words *agent* and *agency* may recall images of individuals sitting in featureless buildings filled with desks, dutifully helping others. They are often associated with an administrative unit or some kind of business enterprise: travel agent, insurance agent, advertising agent, intelligence agent. At a time that calls for transformative changes to secure the future for humans and other species, we need an active and vibrant way to capture the potential for people to make a positive difference. The word ildsjel fulfills this need.

FIRE SOULS

An ildsjel is literally a "fire soul." The word communicates both energy and spirit. It conveys a burning force that is powerful and productive, a force ready to spread like wildfire, making way for new seeds to germinate. Ildsjel describes a lively spirit that is aware, engaged, and ready to act. Geelmuyden points out that many ildsjeler have managed to transmute anger into a moral outrage that compels them to act. Ildsjel or "fire soul" adds human vitality to the concept of *agent*, which is typically assigned a rational, calculated approach to change.

Vitality is a power or energy that is present in all living things. Vitalism holds that life depends on an elán vital, or a nonmaterial life force that does not rely on purely chemical or physical forces. Indeed, it is sometimes used as an explanation for the spirit or soul, since it is not observable and can only be validated through experience or consciousness.[12] Over the last century, research in genetics and biology has led to the dismissal of vitalism as a philosophy. However, scholars in the environmental humanities and social sciences have recently reengaged with vitalism, with some even arguing that there is a vibrant life force in all material things. For example, political theorist Jane Bennett describes a materialism that, in contrast to a mechanistic or deterministic materialism, recognizes matter as "vibrant, vital, energetic, lively, quivering, vibratory, evanescent, and effluescent."[13] She does not reduce political agency to human agency, instead recognizing a vibrant, nonhuman agency that is inextricably enmeshed with human culture.

Ildsjeler often engage with transformative change within a dominant paradigm that is rational, individualistic, and deterministic. This worldview considers humans as separate from nature and separate from each other, since matter is seen as lifeless or dead.[14] Indeed, philosopher Andreas Weber points to the dark side of the Enlightenment habits of thought that have served as the foundation for modern scientific and technological progress: "The Enlightenment project has no use for notions of life, sentience, experience, subjectivity, corporeal embodiment and agency."[15] Adherence to the Enlightenment paradigm can make collective and collaborative change seem impossible or unrealistic, leading to the widespread belief that people cannot and do not make a difference—except by changing their own behaviors and consumption choices. The potential for an individual or group to generate large-scale, positive social and political change is often trivialized or ridiculed.

Weber offers what he calls Enlivenment as an alternative. This concept seeks to "expand our view of what human beings are as embodied subjects," linking rationality with subjectivity and sentience to capture the "aliveness" of a person.[16] Whereas the Enlightenment paradigm recognizes only limited and constrained human agency, the Enlivenment paradigm acknowledges that all humans have the capacity to be ildsjeler, or what leadership theorist Monica Sharma calls "principled game changers who also inspire others to commit to action." This, Sharma explains, involves "a combination of listening deeply, speaking responsibly, and generating new societal conversations drawn from our inner capacities and universal values."[17]

BURNING *FOR* VERSUS BURNING *OUT*

Burning passionately to create positive change without strong economic, institutional, and civic support brings with it the risk of burnout. The notion of burning out seems to recognize that we have a fire within us that needs to be fed and nurtured. Even a single voice of encouragement or a small sign of support can keep the fire burning in an ildsjel. Without this fire, an ildsjel may be reduced to serving as just another rational agent of change. The idea of having a fire soul resonates with what Joan Borysenko, a pioneer in integrative medicine, refers to as an experience of "fire in the soul." She considers this to be a process where one's masks, veils, and facades are painfully burned away to reveal an inner light: "When our souls are on fire, old beliefs and opinions can be consumed, bringing us closer to our essential nature and to the heart of healing. Experiences of inner burning have been called dark nights of the soul."[18] As geographer Lesley Head suggests, such "dark nights" may be necessary for addressing global environmental problems like climate change. In describing the important role of emotions in the Anthropocene, she argues that we need to acknowledge feelings such as grief and denial so as to enact effective politics.[19] This recognition is the first step in discarding the crippling belief that change is not possible and in releasing the vitality of the ildsjel that already exists in all of us.

WORDS MATTER

The words we currently use to describe the role of individuals in processes of social transformations inadequately represent humanity's true potential. This is a problem because words are potent: they transmit both meaning and emotion. How we talk about change influences the possibilities that we imagine, act on, and eventually realize. The word *agency* communicates a limited capacity to inspire collective change. In contrast, an ildsjel is vital, alive, and connected to their community. The word ildsjel adds to our conception of the human potential for transformation by conveying how individuals can and do make a difference. As fire souls, we already have a burning force within us that connects us to others and can activate change.

There is an urgent need to empower people. We need the fire in our souls to be activated by our moral outrage about the current social and environmental conditions and risks. We also need empowering expressions of humanity's potential to contribute to a healthy world in which we can all thrive. Ildsjeler have the capacity to light the fire in others and generate transformations to a more just and sustainable world. The Norwegian word ildsjel is an example of a word that has the potential to activate collaborative change in a way that *agency* cannot.

NOTES

1. Will Steffen, Katherine Richardson, Johan Rockström, et al., "Planetary Boundaries: Guiding Human Development on a Changing Planet," *Science* 347, no. 6223 (2015): 1259855.

2. Nik Brandal, Øivind Bratberg, and Dag Einar Thorsen, *The Nordic Model of Social Democracy* (Basingstoke: Palgrave Macmillan, 2013).

3. Guri Mette Vestby, Frants Gundersen, and Ragnhild Skogheim, *Ildsjeler og lokalt utviklingsarbeid*, NIBR report, 2014, http://www.hioa.no/content/download /109210/2606477/file/NIBR-rapport%202014:2%20(PDF)_afff6970759350d 826794dc5dbf1d117.2014-2.pdf.

4. Niels Christian Geelmuyden, *I hodet på en ildsjel* (Fornebu, Norway: Dinamo Forlag, 2010).

5. Geelmuyden, *I hodet på en ildsjel*, 60–61, our translation.

6. Robert Kegan and Lisa Laskow Lahey, *Immunity to Change: How to Overcome It and Unlock the Potential in Yourself and Your Organization* (Boston, Mass.: Harvard Business Press, 2009).

7. See Simon Dalby, "Framing the Anthropocene: The Good, the Bad and the Ugly," *Anthropocene Review* 3, no. 1 (2016): 33–51.

8. Thomas Dietz, Eugene A. Rosa, and Richard York, "Driving the Human Ecological Footprint," *Frontiers in Ecology and the Environment* 5, no. 1 (2007): 13–18.

9. Eric Bonabeau, "Agent-Based Modeling: Methods and Techniques for Simulating Human Systems," *Proceedings of the National Academy of Sciences of the United States of America* 99, no. 3 (2002): 7280–87.

10. Anthony Giddens, *The Constitution of Society: Outline of the Theory of Structuration* (Berkeley: University of California Press, 1986).

11. Harold Wilhite, *The Political Economy of Low Carbon Transformation: Breaking the Habits of Capitalism* (New York: Routledge, 2016).

12. Alexander Wendt, *Quantum Mind and Social Science: Unifying Physical and Social Ontology* (Cambridge: Cambridge University Press, 2015).

13. Jane Bennett, *Vibrant Matter: A Political Ecology of Things* (Durham, N.C.: Duke University Press, 2010), 112.

14. See Wendt, *Quantum Mind.*

15. Andreas Weber, *Enlivenment: Towards a Fundamental Shift in the Concepts of Nature, Culture and Politics* (Berlin: Heinrich-Böll-Stiftung, 2013), 15.

16. Weber, *Enlivenment*, 15.

17. Monica Sharma, *Radical Transformative Leadership: Strategic Action for Change Agents* (Berkeley, Calif.: North Atlantic Books, 2017), xvi.

18. Joan Borysenko, *Fire in the Soul: A New Psychology of Spiritual Optimism* (New York: Grand Central, 1994), 4.

19. Lesley Head, *Hope and Grief in the Anthropocene: Re-conceptualising Human–Nature Relations* (New York: Routledge, 2016).

In Lak'ech—A la K'in

John Esposito

Pronunciation: een lak'ech—a la k'een
(in lak:ɛch—a la k:in)

Part of Speech: Salutation

Provenance: Ancient Maya (Yucatec)

Example: Upon seeing her neighbor, a Maya woman says in lak'ech, which elicits the reply a la k'in. In lak'ech—a la k'in; in lak'ech—a la k'in.

On a remote mountainside in Mesoamerica, two Maya meet, and, as their people have for many centuries, they exchange the following words: in lak'ech—a la k'in. This ancient Maya greeting can be rendered into English as "I'm another you," to which a respondent would utter, "You're another me." Whereas salutations such as "hello" typically function as a formulaic means of attracting attention, in lak'ech—a la k'in affirms an irrefutable bond between interlocutors that is at once psychological and spiritual. The everyday repetition of the Maya greeting might be one way of instilling a greater sense of interdependence because the way we address each other has a powerful effect on all that follows. It has the potential, moreover, to counter the dualistic assumptions that have contributed to the advent of the current geologic age known as the Anthropocene.

Greetings are an essential component of initial encounters that include nonverbal signals designed to confirm recognition. As with *hola* in Spanish, simply acknowledging another's presence is the primary function of "hello" or "hi" in English. The French *bonjour* and the Chinese *nihao* go a step further, conveying good wishes to the addressee. In Russian, *zdravstvuyte* exhorts its interlocutors to be healthy. The rather poetic *as-salām-alē-kum—wa alē-kum as-salām* (peace be upon you—and also upon you) helps establish an auspicious platform upon which Arabic speakers can converse.[1] Greetings also serve to identify and classify individuals into meaningful categories, often depending on the sociocultural context.[2] Accordingly, they may affirm asymmetric relations while highlighting the relative status of participants.[3] What most greetings appear to have in common is that they reinforce a self–other dichotomy that persists within a particular speech community. To the degree that these sentiments amplify seemingly ingrained differences, they, unlike in lak'ech—a la k'in, help instill a sense of separation that is a defining feature of anthropocentrism.[4]

To establish in lak'ech—a la k'in as a greeting in the current global lingua franca, English speakers must be convinced that this neologism is not a replacement for the quotidian "hello" but its complement. The two have distinct yet overlapping functions. The latter is a salutation designed to attract someone's attention or acknowledge his or her presence, whereas the ancient Maya greeting goes a step further to establish a shared field of meaningful interaction. Yet the transliteration of in lak'ech—a la k'in into English might make for a rather harsh string of sounds. The glottalized consonants that distinguish meaning in Maya are, according to the linguist John Montgomery, difficult for nonnative speakers to reproduce.[5] (The glottal stop is indicated by an apostrophe after the consonant.) Therefore, rather than follow "hello" or "hi" with "How are you?"—a phatic query that often engenders fabrication—the suggestion here is to utter the Maya greeting's translation: "I'm another you." This elicits the reply "You're another me." In effect, "I'm another you" would not replace hello but reinstate the latter's linguistic boundaries as a mere means of attracting attention, thereby opening discursive space for adoption of the former.

The sense of interdependence at the core of "I'm another you" stands in stark contrast to anthropocentrism. Although it has come to the fore over the past two centuries as an epistemological corollary to the modern industrial era, philosopher Max Oelschlaeger traces the roots of anthropocentrism to the advent of the Neolithic Age (circa 10,000 BCE), when the idea that humans are somehow distinct from the rest of creation began to take hold.[6] The sedentary lifestyle precipitated by the agricultural revolution might have facilitated a closer connection to place, yet it led to a gradual sense of estrangement from the nonhuman world. Domestication of plants and animals for food is the quintessential example of humans' attempt to manage and control nature by manipulating other life-forms. Manipulation is best achieved when the object under control is effectively delinked not only from its environment but also from its exploiter. This is the crux of the scientific method, which is based on fundamental divides between mind and matter, subject and object. Sedentariness also hastened population growth, thereby compounding the need to appropriate more resources to support a single species. An exponential increase in numbers has impelled humans to occupy what were previously inhospitable regions, which has contributed to the myriad ecological ills we are currently witnessing, such as habitat destruction and species extinction.

The looming water crisis is arguably the most significant consequence of an overreliance on an anthropocentric worldview. Often treated as a waste repository (consider words such as *drain, effluent, sewer,* and *cesspool*), large amounts of water have become unfit for human consumption. The overpumping of wells and aquifers causes arable land to lie fallow. This will eventually exacerbate food shortages while increasing the volatility of global grain and commodities markets, leaving the most vulnerable at risk of hunger. In less than a decade, a majority of the global population is expected to be living under such conditions.[7] The usual response, couched in the detached rhetoric of better management, mitigation efforts, and policy changes, proceeds from the assumption that water is a renewable resource that can be effectively monitored and distributed. At the height of this folly is the bottled water industry's attempt to position itself as a buffer against inadequate public services or diminishing supplies. Water, however, is not something that can be readily controlled; as the Maya evidently understood, it constitutes a cycle of which humans are only one part.

Ecocentrism offers an epistemological challenge to anthropocentrism by ascribing value to complex systems as well as the vital relations that bind them together. Rather than seeking to manipulate and control, the overriding goal is to empathize and connect with the nonhuman world in a way that recognizes its innate value. Indeed, all entities (organisms, species, and systems) are valued not because of some intrinsic quality each may possess but because they are integral to the existence of others with which they interrelate. Once relationships are held to be primary, the perspective of each organism, species, or system is thus afforded equal standing.[8] What ecocentrism ultimately entails, therefore, is a radical decentering, which paradoxically occurs when it is acknowledged that there are myriad centers, only one of which is occupied by human beings. Each center or node is connected to all others in a vast web of interrelations. Ecocentrism, properly construed, subsumes humanity within a larger whole.

In lak'ech—a la k'in provides an ideal entryway into thinking from such a perspective.[9] It involves a reciprocal exchange between equals that elicits the knowledge that giving and receiving are inseparable, that what is perceived as difference is not incompatible with oneness, which elides all distinctions at a deeper level of consciousness. The Maya greeting thus has the power to transcend tribal, linguistic, or religious affiliations. Its use affirms existential solidarity and coexistence—an enduring perspective that calls into question cultural conventions that reinforce division and separation. In lak'ech reminds participants that they share a spiritual affiliation that supersedes the spatial and temporal constraints of personal identities. The potential psychological implications for the individual are profound: in lak'ech—a la k'in suggests that everyone is part of something greater. When cooperation is understood as a form of self-preservation, for instance, empathy is automatic—not just to human beings but to all species in the web of life.[10]

Aside from the radical implications of adopting in lak'ech—a la k'in, the Maya present us with a moral lesson on the hazards of foregoing a lifestyle deeply rooted in a holistic worldview. Maya civilization flourished for millennia in a geographically diverse and climatically volatile region because of a fundamental belief in the vital essence of all sentient beings, including stones and streams, existing in dynamic cycles of interdependence.[11] This integrated outlook informed their agroecological subsistence practices, such as intercropping and fallowing.[12] Archeological findings provide evidence of the way they maintained a sufficient supply of clean water, despite a prolonged dry season and frequent droughts, by creating reservoirs that mimic the plant–insect balance responsible for purifying wetlands.[13] While living in relative harmony with their nonhuman neighbors, the Maya also utilized terracing, forest gardens, and raised fields as a means of satisfying their everyday needs.[14] A symbiotic lifestyle wedded to natural cycles and processes allowed their culture to flourish until the pursuit of material wealth and fame drove an aristocratic elite into a debilitating power struggle.[15] Deforestation and soil erosion were the most significant consequences of internecine strife that, coupled with an extended period of severe drought, led to the collapse of their remarkable civilization.[16] The central message here is that the sense of reciprocity, coevolution, coexistence, and empathy engendered by the traditional Maya greeting must be extended to all beings if a sustainable future is to be realized.[17]

How might this ancient Maya greeting be adopted into English? The Maya themselves are arguably best positioned to act as ecoambassadors to promote the virtues of in lak'ech—a la k'in, especially the thousands of native speakers currently residing in the United States. They are part of a blossoming pan-Maya movement that seeks to protect their cultural and linguistic heritage.[18] Awareness of the rise and fall of Maya civilization, as succinctly delineated by historian Clive Ponting, for example, can be accompanied by the introduction of their traditional greeting.[19] In an electronically interconnected, social media–savvy era where neologisms are coined and communicated at an unprecedented speed, this would not be without precedent. The challenge is how to make a few simple words prominent against a ubiquitous background of incessant data and information. One way would be to persuade influential groups or organizations, such as the United Nations or its affiliates, to use the Maya greeting either as a motto or in promotional materials. Another approach could be to encourage activist communities to adopt in lak'ech, or "I'm another you," as a common greeting in their communications. This seems particularly relevant to the thousands of groups worldwide who are currently working to facilitate the transition to an ecocentric age.[20]

The undeniable advantage of this loanword is that greetings are a universal feature of human interaction; they are enunciated countless times every day. Although they are not commonly used to convey information, nor are they closely linked to identity, in lak'ech (I'm another you) enables both. It arouses recognition and confirms connection while instilling an essential truth: that earthly existence is an interconnected web of reciprocal relationships. Its frequent repetition can thus function as a collective mantra for a species on the cusp of an existential transformation.[21] Its utterance could help to counteract the debitating effects of anthropocentrism. Each iteration of this rhythmic phrase would represent a brief respite from the stress of living in an increasingly alienated age. Indeed, whether it ultimately acts as a spiritual supplement or communicative complement, it is time to introduce in lak'ech—a la k'in as an enduring ritual that honors its speakers as well as the irrefutable bonds between them.

NOTES

1. The six official languages of the United Nations, used as a first or second language by nearly half the world's people, are English, Spanish, French, Chinese, Russian, and Arabic.

2. Alessandro Duranti, "Universal and Culture-Specific Properties of Greetings," *Journal of Linguistic Anthropology* 7, no. 1 (1997): 63–97.

3. See Dele Femi Akindele, "Lumela/Lumela: A Socio-pragmatic Analysis of Sesotho Greetings," *Nordic Journal of African Studies* 16, no. 1 (2007): 1–17; Abdulai Salifu Asuro and Ibrahim James Gurindow M-minibo, "Convergence and Divergence Strategies in Greetings and Leave Taking: A View from the Dagba Kingdom in Ghana," *International Journal of Linguistics* 6, no. 4 (2014): 224–37; and Peter G. Emery, "Greeting, Congratulating and Commiserating in Omani Arabic," *Language, Culture, and Curriculum* 13, no. 2 (2000): 196–216.

4. Anthropocentrism entails not only a detachment from other life-forms but also, and more importantly, a privileging of human concerns. In the venerable chain of being, humans are at the planetary apex. From such an exalted perch, they presume the utilitarian right to preside over earthly life. The self–other divide that is evident in typical greetings is thus extended to the treatment of other species with nature as the ultimate other.

5. See John Montgomery, *Maya–English: English–Maya (Yucatec) Dictionary and Phrasebook* (New York: Hippocrene Books, 2004).

6. See Max Oelschlaeger, *The Idea of Wilderness: From Prehistory to the Age of Ecology* (New Haven, Conn.: Yale University Press, 1991).

7. UN-Water, "A Post-2015 Global Goal for Water," 2014, http://www.zaragoza.es /contenidos/medioambiente/onu/1090-eng_A_Post-2015_Global_Goal_for _Water.pdf.

8. This is neither entirely achievable nor desirable from an ecocentric perspective, for it can only be attained if humans and nature are in fact separate entities to begin with; conversely, if the human and natural worlds are not distinct, then the positing of intrinsic value to nature occasions contradiction.

9. The parallelism in the construction of in lak'ech—a la k'in is a prominent feature of Maya conversation. Brown shows how this dialogic repetition of mostly the same content with a few minor syntactic alterations (concerning, for example, modifiers or connectives) constitutes a collaborative style of oral interaction that children learn to emulate. See Penelope Brown, "Conversational Structure and Language Acquisition: The Role of Repetition in Tzeltal," *Journal of Linguistic Anthropology* 8, no. 2 (2000): 197–221.

10. As with other species, intimate greetings have the power to forge essential bonds, foster social cohesion, and maintain relationships based on tolerance and cooperation. See, for example, Jennifer E. Smith, Katherine S. Powning, Stephanie E. Dawes, et al., "Greetings Promote Cooperation and Reinforce Social Bonds among Spotted Hyaenas," *Animal Behaviour* 81, no. 2 (2011): 401–15; and Jessica C. Witham and Dario Maestripieri, "Primate Rituals: The Function of Greetings between Male Guinea Baboons," *Ethology* 109, no. 10 (2003): 847–59.

11. See Michael D. Coe and Stephan Houston, *The Maya*, 9th ed. (London: Thames & Hudson, 2015).

12. See Anabel Ford and Ronald Nigh, *The Maya Forest Garden: Eight Millennia of Sustainable Cultivation of the Tropical Woodlands* (London: Routledge, 2015).

13. Lisa J. Lucero, Joel D. Gunn, and Vernon L. Scarborough, "Climate Change and Classic Maya Water Management," *Water* 3, no. 2 (2011): 479–94.

14. See Walter R. T. Witschey, "Subsistence," in *Encyclopedia of the Ancient Maya*, ed. Walter R. T. Witschey (Lanham, Md.: Rowman & Littlefield, 2016), 321–23. For a detailed study of contemporary Maya agricultural practices, see Narciso Barrera-Bassols and Victor Manuel Toledo, "Ethnoecology of the Yucatec Maya: Symbolism, Knowledge and Management of Natural Resources," *Journal of Latin American Geography* 4, no. 1 (2005): 9–41.

15. See Billie L. Turner and Jeremy A. Sabloff, "Classic Period Collapse of the Central Maya Lowlands: Insights about Human–Environment Relationships for Sustainability," *Proceedings of the National Academy of Sciences of the United States of America* 109, no. 35 (2012): 13908–14.

16. See Nicholas P. Dunning, Timothy P. Beach, and Sheryl Luzzadder-Beach, "Kax and Kol: Collapse and Resilience in Lowland Maya Civilization," *Proceedings of the National Academy of Sciences of the United States of America* 109, no. 10 (2012): 3652–57.

17. Maya history might have turned out quite differently if not for the crater that was produced by the six-mile-wide meteor that struck the Yucatán Peninsula sixty-six million years ago. The meteor's impact—referred to as an extinction event—created numerous sinkholes that became a major source of drinking water for the first human settlements in what is known as the Maya lowlands.

18. See Coe and Houston, *Maya*.

19. See Clive Ponting, *A New Green History of the World: The Environment and the Collapse of Great Civilizations* (New York: Penguin, 2007).

20. See Paul Hawken, *Blessed Unrest: How the Largest Social Movement in History Is Restoring Grace, Justice, and Beauty to the World* (New York: Penguin Books, 2007).

21. The mantric qualities of "I'm another you"—"you're another me" include the inverse ordering of soft sounds, and two sets of syllables (1-3-1) in a reciprocal relationship that constitute a 1-3-1–1-3-1 prosodic pattern. Moreover, the initial "I'm" is analogous to "Om," which is considered by Hindus and Buddhists to be the elemental sound representing universal consciousness.

Pronunciation: me-da-hyoo-man

(mɛtəːhjuːmən)

Part of Speech: Noun

Provenance: Superhero comic books, film, and television

Example: Converting a landfill into a park is truly a metahuman achievement!

A glowing ring descends from the sky. It beckons: "You have the ability to overcome great fear. You are chosen."

A letter arrives by owl: "We are pleased to inform you that you have been accepted at Hogwarts School of Witchcraft and Wizardry."

A door into danger opens: "Welcome to the X-Men! Hope you survive the experience."

Readers have long imagined themselves into such encounters, taken here from *Green Lantern*, the Potterverse, and Marvel Comics, but common to many fictional realms. On the surface, these tableaux are vehicles of adolescent wish-fulfillment, summoning us into adventure beyond childhood. Lurking below, however, are traditional rites of transformation—prophetic commissioning, artistic epiphany, and scientific discovery—and their institutional equivalents—joining the assembly, founding a movement, and advancing higher learning. By extending these traditions, contemporary media transmit the modernist revival of art, ritual, and myth meant to challenge and enlighten us.[1] This is not an evil science project or a plan for world domination;

popular cultures evolve by figurative means the kinds of people who can literally defend the world. Like the pulse of a Green Lantern or a witch's spell, tales of the fantastic shape the creatures and communities we need to become. We have only to heed the call! Enter the metahuman.

In comic books, film, television, and video games, a metahuman, or meta, is a person with extraordinary abilities (*meta*, "beyond" + human)—hero, villain, or otherwise. Popularized in the Pixar film *The Incredibles* (2004), the term originates in the DC Comics limited series *Invasion!*, published in 1988. Though the term metahuman now denotes anyone with powers, *Invasion!* emphasizes the origin of these abilities in genetic variation, revealing the influence of an earlier term, *mutant*, from Marvel Comics. As book 1 opens, an alien race called the Dominators expose kidnapped humans to physical threats that would destroy normal people. Only seven test subjects survive, and the Dominators conclude that these survivors possess latent superpowers. This potential is considered a threat to Dominator hegemony. Our planet "is apparently capable of generating a dazzling array of heroes possessed of powers as unpredictable as they are dangerous."[2] This danger prompts the invasion, which aims to destroy the genetic capacity to develop superpowers, known as the metagenome. *Invasion!* and its legacy shifted the discourse of power in Anglophone popular culture beyond the realm of comics. While Marvel had been publishing stories about genetic variation since the 1960s in its iconic X-Men titles, DC heroes such as Wonder Woman, the Flash, and Superman gained their abilities from sources other than mutation. With its turn to the genome as the origin of superpowers, DC emphasized the ability of humanity to change for the better. The planet was in peril, and humans naturally developed the powers to save it.

Yet capability is not enough. We must also choose to defend the world. The ethical commitment to world defense requires a culture of metahumanism. In the absence of a community of practice, we find the aloof Übermensch—or worse, the villain. As Ramzi Fawaz explains, "What distinguished the superhero from the merely superhuman . . . was its articulation of an extraordinary body to an ethical responsibility to use one's power in service to the wider community."[3] Therefore, metahumanism requires a process of learning to evoke, discipline, and apply one's powers to the common good. Like the X-Men, we have to cultivate mutant abilities in order to make a better world. Metahumanist discourse democratizes the aristocratic concept of the Übermensch and negates eugenic typologies of the superior breed threatened by criminal, insane, and degenerate forces. This would not be the first time that a threatened population had imagined superhuman mobilization. Many histories of the superhero note that Superman, Captain America, and other Golden Age heroes were created by Jewish Americans as ironic reversals of the Aryan overman.[4] Steve Rogers and Clark Kent may look like Nazi ideals, but they are enemies of fascism. In the idea of the metahuman, the irony untwists into a democracy of virtue: now a hero can look like anyone. "What if," the Dominators muse, "it is discovered that this genetic trait is inherent not just in *some* humans, but *all*?"[5] In that case, the power to protect the world waits in everyone, and it is the function of metahumanism to cultivate our capacity to defend the Earth, even from our former selves and their planetary systems of doom. Herein, I suggest, lies the potential of metahumanism to inform the project of constructing ecotopia in the twenty-first century.

Metahumanism is necessary because of the effects of the Anthropocene, the term proposed for the geologic era in which human activity comes to radically alter the lithosphere, hydrosphere, and atmosphere.[6] Environmental humanists and activists have adapted this term as a tool of critique. As Christophe Bonneuil and Jean-Baptiste Fressoz explain, "the Anthropocene is a sign of our power, but also of our impotence" because it is marked by a damaged atmosphere, disrupted climate, mass extinction, and violence against the poor.[7] Though their book is called *The Shock of the Anthropocene*, they insist that "we should not

act as astonished ingénues who suddenly discover they are transforming the planet": the project of the elites of the Global North has been
the mastery of nature throughout modernity.[8] The name of the epoch
forces us to ask, "What to do now?"[9] Critics have already taken the
term to task for ascribing the problem to Anthropos ("man" as a species) without contending with the systems of power—global capitalism,
colonialism, patriarchy—that drive ecological degradation.[10] However,
they go beyond critique to offer new names for other ages that struggle
to emerge. My favorite is the Phronocene, the age of practical wisdom
(from the Greek word *phronesis*, "intelligence" or "wisdom"), which
would mark a disruption of the disruption of Earth systems, an attempt
to live wisely with other species and elements.[11] The Phronocene would
be a chastened age of ecotopia—not a perfected world, but a homeplace
honored with genuine care.[12] This chronotope fits Donna Haraway's
project of "staying with the trouble"—neither repressing awareness
of the crisis nor despairing of effective action.[13] The question arises,
then, of how to move beyond adjustments in individual consumption—
imagined as governmental and personal austerity—to evolve cultures
capable of staying with the trouble in order to cultivate planetary flourishing. Metahumanism would help us to move beyond tropes of the
Anthropos as supervillain, the Anthropocene as world domination, and
academic humanism as sidekick to the Big Bad. As I argue elsewhere,
nerd culture offers tropes, narratives, and ethics that ground metahumanist philosophy and practice.[14] Here I offer two visions drawn from
American literature and television: Octavia E. Butler's Xenogenesis trilogy (1987–89) and showrunner Greg Berlanti's TV version of *The Flash*
(2015).[15]

Though Butler's Xenogenesis trilogy is not an ecotopian work, it
fits the parameters of what scholars call a critical utopia, which Ildney
Cavalcanti argues displays a consciousness of utopian tropes and offers
a revision of the "ideal place" through the integration of dystopian elements.[16] In the opening book of the trilogy, *Dawn*, Butler tells the story

of Lilith Iyapo, an African American woman who wakes up on an alien spaceship hundreds of years after a nuclear war.[17] While she and other survivors slept, the Oankali—a race of tentacled gene traders with three sexes and a strong sense of family—restored the Earth to a semblance of ecological health. They offer to send humans to it, as long as we agree to interbreed with the Oankali to produce a hybrid race that will inherit the Earth. Though critics emphasize the overtones of slavery, colonialism, and genocide in the novel, Lilith is a moral agent driven by her consciousness of these histories. Lilith can neither escape the Oankali nor deny the near destruction of her world by nuclear war, and the Oankali forbid human self-rule on Earth. Having failed to escape, she chooses to mentor the new hybrid species, which inherits human intelligence without the drive for domination that the Oankali consider humanity's fatal flaw. Lilith's choice preserves a human legacy under threat of annihilation, but it also reflects the will to restore other species and the biosphere itself. Lilith is thus an exemplar of first-generation metahumanism: a teacher who ensures the continuation of history by choosing to shift shape, hoping that her students will be better but knowing that they will be different. The next generation gains metahuman powers such as rapid healing and great strength, but their most important power is an absence: they have no instinct for social hierarchy and therefore no urge to fight for dominance. Butler's critical ecotopia wrestles with the drive to be free, to reject Oankali compulsion, in light of species preservation and planetary health. It offers metahumanists the knowledge that change depends on the power to relinquish dominion in favor of alien pleasures.

If the Xenogenesis trilogy finally grants the status of world-historical figure to a woman of color, the TV series *The Flash* moves the white savior into a matrix of multiracial community. In its current incarnation on the CW network, *The Flash* enacts a standard comic book origin narrative: young Barry Allen is caught in an explosion that grants him superhuman speed and resilience. In his classic form, the Flash is a modern Hermes, the youngest and most lighthearted of heroes. In bringing him to television, producer Greg Berlanti poses the questions: How can anyone be wise enough to use superpowers ethically? How can virtue and good judgment team up for the common good? Berlanti answers the first question by surrounding the Flash with family, friends, and colleagues who mediate his metamorphosis from insecure youth to responsible adulthood. Even his opponents teach him to be a better meta. Season 1 makes clear that it is impossible for an isolated individual to use his powers rightly: the greater hero is the extended community. The Flash is really Team Flash, grounded in its beloved Central City. Seasons 2 and 3 extend the logic of metahuman praxis into that community. As they help Barry develop, others discover their own gifts.[18] Some of these gifts are less paranormal than exemplary: Iris West, who in previous iterations of the Flash's franchise was the girlfriend, discovers that she needs no catastrophe to manifest the intelligence, cunning, and bravery of an investigative journalist. The cultivation of individual gifts strengthens the team's network of virtue and wisdom, which helps the city to flourish. In this framework, the hero is the lightning that catalyzes a metahuman polity.

Given the metahumanist focus on transformation, it is no wonder that the Xenogenesis trilogy and *The Flash* draw from the tradition of the bildungsroman, the novel of maturation, self-cultivation, and entry into adult society. For the same reasons, both feature pedagogical subplots in which students grow up to be mentors themselves. This is useful for a practical ecotopianism because it answers the vexing question of how to get there from here: education must be one vector of ecotopian culture. This is more than theoretical, however. There are already pedagogical projects that feature metahumanist principles. One example of pedagogical metahumanism is *Operation Superpower*, a participatory chamber opera founded by a composer, two baritones, and a pianist. It began touring in 2013 for K–12 audiences at schools, theaters, and opera houses in the United States and Canada. The premise of the project is that "a SUPERPOWER is a student's inner talent and a SUPERHERO is one who uses that talent to help others!"[19] Structured loosely around the origin stories of an alien orphan (Superman) and an urban vigilante (Batman), the performance asks student audiences to respond to the musicians with demonstrations of their own powers, artistic or otherwise. This structure creates the reciprocity that the project asks teachers to extend after the performance is over. To provide a framework for local pedagogy, the performers offer theory and praxis: icebreakers and vocal warm-ups followed by an exposition of the heroic virtues of courage, hope, imagination, honesty, and friendship. This is an example of Aristotle's theory that virtues can be cultivated through the formation of habits by wise action toward a good end. It is his *Nicomachean Ethics* 101 transformed for a democratic polity in the pragmatic philosophy of John Dewey. This system is no longer for the propertied citizen-in-training male subject alone but for everyone willing to commit to its principles. The results thus far have been both adorable and effective: teachers report "transfer," the successful application of skills from one context to another. Skills transfer through cognitive and metacognitive activities. Students apply the show's logic to other activities, reflect on their learning, and assess their achievement as part of a team. The project provides material anchors such as stickers, handouts, and T-shirt patterns to turn initial enthusiasm into long-term retention. Because the show requires only three performers, it is easy to create multiple casts that are inclusive in terms of gender and race. Moreover, because of its initial success, *Operation Superpower* is moving ahead with plans for a full-scale opera. In only a few years, this grassroots effort led by performing arts students has demonstrated that metahumanism can be leveraged to artistic and pedagogical ends.

If *Operation Superpower* confirms the pedagogical potential of metahumanism, then the fan activism of the Harry Potter Alliance[20] demonstrates its political potential.[21] Since 2005, the HPA has turned "fans into heroes" by organizing local groups dedicated to the magical world of J. K. Rowling into political task forces. The founder, Andrew Slack, characterizes the group's strategy as "cultural acupuncture," the creation of a better world by focusing fan energy toward particular goals.[22] The HPA has over 100,000 members in seventy countries; given these numbers, the HPA provides a strong counterexample to the traditional suspicion of popular culture shared by conservative commentators such as Allan Bloom and Marxist critics of the Frankfurt School. Far from seducing fans into fascistic groupthink, HPA supports diverse projects ranging from charity work to political activism, as documented by media scholar Henry Jenkins:

> The group collaborates with more traditional activist and charity organizations, such as Doctors for Health, Mass Equity, Free Press, The Gay-Straight Alliance, and Wal-Mart Watch. When the HPA takes action, the results can be staggering: for instance, it raised $123,000 to fund five cargo planes transporting medical supplies to Haiti after the earthquake. Its Accio Books! Campaign has collected over 55,000 books for communities around the world. HPA members called 3,597 residents of Maine in just one day, encouraging them to vote against Proposition 9, which would deny equal marriage rights to gay and lesbian couples. Wizard Rock the Vote registered more than a thousand voters.[23]

The HPA is significant from a metahumanist perspective for reasons that transcend its results. As a group that operates through the tropes of metamorphosis and heroism, it shows that the rhetoric of virtue transcends the superhero framework, appealing to a different fandom cathected to another kind of story world. In the Harry Potter series, magical ability runs in families, but it also appears in nonmagical Muggle families as well. Here, the eugenic framework of the comic book world is destabilized by the political debate about blood heritage, with racist "pureblood" wizards allying with the villain, Voldemort, to deny the rights of "mudbloods," a pejorative term for wizards with non-magical lineage. Because political debates about racism, misogyny, slavery, and fascism structure the story world itself, the HPA does not need to connect Rowling's work to contemporary politics so much as extend its politics into the world of the audience. The HPA further translates story structures such as the sorting of students into academic houses and the competition between houses into organizational and motivational principles. Members are sorted by their interests into houses that mirror the Hogwarts School of Witchcraft and Wizardry, and houses take on particular projects, competing with others to maximize results. Like *Operation Superpower*, the HPA translates the capacity to imagine a better self and world into the practice of fostering utopia.

These seeds of transformation illustrate how nerd culture can be yoked to civic responsibility, but it remains to be connected to ecotopian politics. Metahumanism offers a direction for humanism foreseen by philosopher Val Plumwood when she called for the development of "environmental culture" as the antidote to human domination of nature. Plumwood argues that "the problem is not primarily about more knowledge or technology, it is about developing an environmental culture that values and fully acknowledges the non-human sphere and our dependency on it."[24] The human sciences are good at culture. As the humanities are currently configured, however, transmission of past culture trumps the creation of culture, though movements of critical making and postcritique are challenging this paradigm.[25] The recent constellation of the environmental humanities moves toward the latter, but it remains framed in terms of high culture. As a scion of popular culture, metahumanism should build on humanistic strategies for the reproduction of canons, the creation of new works, and the critical analysis of culture and move into a constructive mode in order to cultivate ecotopian ways of life. At a moment when the neoliberal university views the arts and humanities as handmaidens of the STEM disciplines, metahumanists should seize the high ground as builders of the Phronocene. As the HPA website notes, "We believe that unironic enthusiasm is a renewable resource." Enthusiasm for environmental justice steers a path beyond a defense of human supremacy and the quest for disembodied immortality, whether technological or religious. Such unabashed utopianism will be difficult, given that the globe has been dominated by realpolitik. Yet it is precisely this fetish for realism that has led to ecocide: the present order is the natural order, and resistance is futile. This is the way Dominators think, which clarifies the cause of metahumanism: to overcome static notions of human nature that prevent us from turning aside our doom.

Against this opposition between *is* and *ought*, we can assert the connection between them as theorized by J. R. R. Tolkien in his essay "On Fairy-Stories," in which he defends fantasy against the charge of escapism: "Why should a man be scorned if, finding himself in prison, he tries to get out and go home? Or, when he cannot do so, he thinks and talks about other topics than jailers and prison-walls? The world outside has not become less real because the prisoner cannot see it."[26] Tolkien views imaginary worlds as places of refuge and reflection, from which one returns to the present world with the capacity to change it. As the HPA website puts it, "We know fantasy is not only an escape from our world, but a means to go deeper into it." For this reason, metahumanism at large should refine its techniques for strengthening the gifts of community members in service of an ecotopian order. Meanwhile, metahumanists in the academy should further this goal through the evolution of assignments, curricula, requirements, and degree programs. By focusing the will to mutate into the arts of change, metahumanists can combine forces to cultivate a ground for ecotopia.

NOTES

My thanks to my colleagues in the Juilliard JAM writing group—especially Lisa Andersen, Aaron Jaffe, Greta Berman, Cory Owen, Fred Fehleisen, and Jordan Stokes—for trenchant feedback.

1. For an account of Anglophone modernism's appropriation of myth and ritual through the concept of "culture," see Marc Manganaro, *Myth, Rhetoric, and the Voice of Authority: A Critique of Frazer, Eliot, Frye, and Campbell* (New Haven, Conn.: Yale University Press, 1992).

2. Keith Giffen, *Invasion!* (Burbank, Calif.: DC Comics, 2016), 13.

3. Ramzi Fawaz, *The New Mutants: Superheroes and the Radical Imagination of American Comics* (New York: New York University Press, 2016), 6.

4. Michael Chabon's novel *The Amazing Adventures of Kavalier and Clay* (2000) connects Superman to the Golem of Prague, the protector of the Jewish ghetto. For a more detailed account by comics industry creators, see Danny Fingeroth and Stan Lee, *Disguised as Clark Kent: Jews, Comics, and the Creation of the Superhero* (New York: Continuum, 2007). Famously, Captain America punches Hitler on the cover of *Captain America* #1 (1941) a year before the United States entered World War II, well in advance of public sentiment in favor of the war. This legacy was recently extended when America Chavez, Marvel's queer Latina hero, time travels to that iconic moment and punches Hitler herself in *America* #1 (2016).

5. Giffen, *Invasion!*, 14.

6. Indeed, one might argue that the origins of "Silver Age" heroes such as the Fantastic Four and the X-Men in radiation exposure marks the dawning of contemporary environmental consciousness as much as Rachel Carson's *Silent Spring*, in which pesticides are represented through implicit metaphor as radioactive fallout.

7. Christophe Bonneuil and Jean-Baptiste Fressoz, *The Shock of the Anthropocene* (New York: Verso, 2016), xi.

8. Bonneuil and Fressoz, *Shock of the Anthropocene*, xi.

9. Bonneuil and Fressoz, *Shock of the Anthropocene*, xii.

10. See, for instance, Donna Haraway, "Anthropocene, Capitalocene, Plantationocene, Chthulucene: Making Kin," *Environmental Humanities* 6, no. 1 (2015): 159–65, and Jason W. Moore, *Capitalism in the Web of Life: Ecology and the Accumulation of Capital* (New York: Verso Books, 2015).

11. Bonneuil and Fressoz, *Shock of the Anthropocene*, 129.

12. The standard conception of *phronesis* enters Greek philosophy with Aristotle's *Nicomachean Ethics*, and in that work, it is a virtue possessed by individuals, not society as a whole. So the idea of a collective wisdom through which societies act prudentially in the face of ecocide would itself be a matter of cultural evolution, at least in terms of canonical Western philosophy.

13. Donna J. Haraway, *Staying with the Trouble: Making Kin in the Chthulucene* (Durham, N.C.: Duke University Press, 2016).

14. Anthony Lioi, *Nerd Ecology: Defending the Earth with Unpopular Culture* (New York: Bloomsbury Academic, 2016), 197–206.

15. The Xenogenesis trilogy was originally published as three separate novels: *Dawn* (1987), *Adulthood Rites* (1988), and *Imago* (1989). An omnibus edition was published as Octavia E. Butler, *Xenogenesis: Dawn, Imago, Adulthood Rites* (New York: Warner, 1989).

16. Ildney Cavalcanti, "The Writing of Utopia and the Feminist Critical Dystopia: Suzy McKee Charnas's Holdfast Series," in *Dark Horizons: Science Fiction and the Dystopian Imagination*, ed. Raffaella Baccolini and Tom Moylan (New York: Routledge, 2004), 47–69.

17. In Jewish folklore and Talmudic commentary on the book of Genesis, Lilith appears as the first wife of Adam, created before Eve to be Adam's peer. When he refuses to treat her as such, she flies over the wall of Eden and into the wilderness, where she gives birth to demons, the *lilim*. The Xenogenesis series can be read as an extrapolation of that tradition.

18. Season 4 features a Flashless Team Flash that must protect Central City in Barry Allen's absence after he disappears into the Speed Force at the end of season 3. This plot represents a new degree of empowerment for the team and also reflects developments in other series in the Arrowverse. *The Flash: The Complete First Season*, DVD (Burbank, Calif.: Warner Home Video, 2015).

19. "Operation Superpower," March 15, 2017, https://www.facebook.com/operation superpower/.

20. https://www.thehpalliance.org/.

21. In 2017, students at Harvard's Kennedy School of Government founded the Resistance School (https://www.resistanceschool.com/#resist), an anti-Trump movement modeled after the student-led resistance to Lord Voldemort and the Death Eaters in *Harry Potter and the Order of the Phoenix* (2003).

22. Henry Jenkins, "'Cultural Acupuncture': Fan Activism and the Harry Potter Alliance," in "Transformative Works and Fan Activism," edited by Henry Jenkins and Sangita Shresthova, special issue, *Transformative Works and Cultures*, no. 10 (2012), https://doi.org/10.3983/twc.2012.0305, ¶ 1.3.

23. Jenkins, "Cultural Acupuncture," ¶ 3.2.

24. Val Plumwood, *Environmental Culture: The Ecological Crisis of Reason* (New York: Routledge, 2001), 3.

25. See Rita Felski, *The Limits of Critique* (Chicago: University of Chicago Press, 2015).

26. J. R. R. Tolkien, "On Fairy-Stories," in *The Monsters and the Critics and Other Essays*, ed. Christopher Tolkien (London: Allen & Unwin, 1983), 156.

Portfolio and Artists' Statements

Blockadia (Ya Basta!)
  Nicolás de Jesús

Untitled 2018 [Dàtóng]
  Rirkrit Tiravanija

Ghurba
  SWOON

Godhuli
  Jonathan Dyck

Heyiya
  Jenny Kendler

Ildsjel
  Lori Damiano

Nahual
  Michelle Kuen Suet Fung

Pachamama
  Yellena James

Plant Time
  Natasha Bowdoin

Water-Wind (Qi)
  Moonassi

Sehnsucht
  Nikki Lindt

Solastalgia
  Kate Shaw

Sueño
  Susa Monteiro

Terragouge
  Maryanto

PLATE 1. Nicolás de Jesús, *Blockadia (Ya Basta!)*

Blockadia (Ya Basta!)

Nicolás de Jesús

En esta Tierra que parece desfalleciente, imposible de soportar el ritmo de sobre explotación de sus entrañas por parte de sus propios hijos; el Ser humano, que idealmente pudiera ser el depositario de la confianza para amarla y conservarla, es lamentable reconocer que está cegado por la ambición generada por su egoísmo, éste se ha convertido en su principal verdugo. Sin reconocer que al atentar en contra de esta naturaleza, se dirige hacia su propia aniquilación y arrastrando a toda especie de vida existente en ella!

Pero acaso no queda una alternativa? Dónde queda la memoria de nuestros antepasados? Por qué no somos capaces de descifrar el mensaje de Amor que heredamos a través de sus luchas históricas alimentadas con nobles ideales hacia los que ahora existimos? Por qué no nos sentimos dignos de respirar el espíritu de fortaleza que heredamos con sus sueños de libertad? Ellos siguen allí con su energía, entrelazados en el poder de la conciencia e incendiar la nuestra! "Levántense y luchen por la vida," nos animan. Los que vienen—sus hijos y nietos—se sentirán orgullosos y honrarán su memoria también. Nos gritan: "No están solos! Estamos todos con ustedes!"

El corazón y la conciencia no pueden fallar, han sido el motor del universo. Griten, lloren y rían hasta sentir la locura para contagiar ese amor al mundo, para romper las cadenas del alma mezquina que les han impuesto los poderosos de este mundo material! Blockadia es el Poder de la Conciencia. Es sentir lo que siente el otro ser humano para entrelazar los brazos del Poder de la Conciencia! Es un Poder del "Ya basta!"

Otro mundo es posible!

On this Earth that is succumbing, unable to withstand the pace of exploitation by its own children, stands a Human Being who ideally could be entrusted with the role to love it and preserve it. It is unfortunate to acknowledge that, blinded by the selfishness generated by its own ambition, he has become its main executioner. Without realizing that by attacking nature, he brings about its annihilation, sweeping away all forms of life present within it!

But is there no alternative left? Where is the memory of our forefathers? Why are we incapable of deciphering the message of love that we inherited from historical struggles fed by noble ideals? Why do we not feel worthy of breathing the spirit of strength that we inherited from their dreams of freedom? They are still there with their energy, interlaced with the power of conscience and setting ours on fire! "Rise and fight for life," they encourage us. Those to come—their children and grandchildren—will be proud and will also honor their memory. They shout to us: "You are not alone! We are with you!"

Our heart and conscience cannot be wrong; they have been the engine of the universe. Shout, cry, and laugh until you feel the madness to spread this love to the world, to break the chains placed by the powerful on your caged soul! Blockadia is the Power of Conscience. It is feeling what another human being feels so as to lock arms with the Power of Conscience! It is the Power of "Enough is Enough!"

Another world is possible!

TRANSLATED BY EDUARDO LAGE-OTERO

In my practice I am often interested in engaging people within frameworks that are preoccupied with what happens when humans come together and share responsibility over simple things. These situations imply a complex relationship to a variety of different concepts and notions of utopia.

The two points regarding the Confucian concept of dàtóng that resonate most with me in Andrew Pendakis's text, which became the foundation for my contribution, are the absence of any consideration of the relationship between humans and nature, on the one hand, and its role as an actual political program during Mao's cultural revolution, on the other.

My sculpture, a lotus flower about to blossom in a Thai military boot sculpted from local clay found near my place in Chiang Mai, is a simple response that combines these two seemingly disparate elements. It is a consideration of the relationship between humans and nature; further, it is a meditation on the tension between utopian ideas and the means of their realization, posing the question of whether dàtóng "remains consigned to the saddest of fates: that of being a beautiful (but largely toothless) idea," or whether it can be much more, ideally beyond and outside of the realm of militarist mass politics.

Rirkrit Tiravanija

Untitled 2018 [Dàtóng]

Ghurba

SWOON

Ghurba conjured images of generative amniotic symmetries. This unfolding of life outward from the center may well be our original home. In creating this piece, I worked from a sense of searching for safety all the way down in the depths, where our very selves were generated.

PLATE 3. SWOON, *Ghurba* >

Malcolm Sen's exposition of godhuli draws together wide-ranging descriptions of light and color—both natural and otherwise—to arrive at "an ethics of place and a metaphysics of possibility." With my piece, I've aimed to mirror Sen's approach, depicting the literal translation of godhuli (cow dust) as a point of departure for an illustration that plays with color, texture, and depth. Here godhuli blurs the distinctions between human and nonhuman entities, both foreground and background, reconstituting them as collections of highlights, midtones, and shadows.

This image depicts godhuli as a redistribution of matter and color, a re-visioning of movement and rest, in line with Alfred Russel Wallace's description of dust as "matter in the wrong place." Without such matter refracting and reflecting the sunshine, he writes, we would be without variations of color, clouds, or rain. Displacement is, in other words, a condition not only for optics but for life. From the footsteps of a herdsman and his cattle to the particles floating through the sky, these small bits of matter reanimate our conceptions of appearance, place, and possibility.

< PLATE 4. Jonathan Dyck, *Godhuli*

Heyiya

Jenny Kendler

At a relatively young age, my mother handed me her worn copy of Ursula K. Le Guin's *The Left Hand of Darkness.* Le Guin (1929–2018) immediately became, and has remained, my favorite author. It is an honor to have been given this wonderful word and concept, heyiya—from her ambitious ecotopian tome, *Always Coming Home*—to illustrate so soon after Le Guin's passing.

In creating a visual evocation of heyiya, I refer to two movements which informed *Always Coming Home*—the world building of utopian science fiction and the back to the earth movement, and their correlating aesthetic styles, sci-fi illustration and 1970s-era counterculture photo collages.

Heyiya is Le Guin's attempt to give form to the living practice of balance, which she both advocated for and demonstrated throughout her writing. Heyiya accepts the dual-natured symmetry of things, darkness alongside light. As she said, "When you light a candle, you also cast a shadow."

Le Guin herself practiced a form of heyiya in her lifelong work to portray complex, nuanced otherness, working against the grain as a female sci-fi writer often depicting women and people of color.

Through the spiraling of heyiya, Le Guin lays out a path to reenchant civilization as a porous social construct not predicated on the domination, human exceptionalism, and extractive model of contemporary capitalism, but guided by acceptance of difference and compassion as well as a living knowledge of who we are as a species.

In my illustration, a human seeks to echo the cosmic in balance with the earthly through an open-handed greeting.

Hand as connector. Hand as seeker. A hinge.

PLATE 5. Jenny Kendler, *Heyiya* >

Ildsjel

Lori Damiano

Upon my introduction to the word ildsjel, a series of faces flashed before me: the ildsjeler in my own life. There was no ambiguity in locating them—they are truly the brightest lanterns and swiftly emerged as a constellation in my mind.

I knew I wanted to paint some kind of human architecture. The ildsjel was to be the foundation linking all of the individuals to each other and grounding them to the earth. After painting a configuration of people embracing in a wreath of mutual support, I made a star encircled by a band of energy. Without consciously realizing it, I had painted a compass. I had envisioned some kind of ring of energy or orbit above the ildsjel, as when a community comes together for a common cause it creates a focused flow of energy that gains momentum at a rate that surpasses our potential as individuals. As Karen O'Brien and Ann Kristin Schorre quote Ingebrigt Steen Jensen as saying, "an ildsjel must get the fire to spread" in order to get enough flow swirling around the community to draw others in.

There is a quiet legacy of ildsjeler who have been able to refocus and redirect us with their actions. Their footsteps illuminate a previously unmarked path on which the rest of us may find better footing, widen our vision, and become more aware of the impact of our actions. In creating this image, I wish to honor the fire spirits of the ildsjel, past, present, and future.

PLATE 6. Lori Damiano, *Ildsjel* >

Michelle Kuen Suet Fung

In Cantonese, the number nine (九) connotes both permanence (久) and dog (狗), the former auspicious and the latter insulting and derogatory. This double play on the number echoes the duality that goes through my mind when I ponder human–animal relationships. Except for a small niche of vegetarians and animal rights activists, most humans have no problem marveling at the wonders of a dolphin's graceful dive at the aquarium, then heading to a steak house.

In this mixed-media drawing, I depict the eye sockets of nine animal groups that the World Wildlife Fund categorizes as critically endangered: rhino (*Diceros bicornis, Rhinoceros sondaicus*, and *Dicerorhinus sumatrensis*), elephant (*Elephas maximus sumatranus*), leopard (*Panthera pardus orientalis*), vaquita (*Phocoena sinus*), saola (*Pseudoryx nghetinhensis*), gorilla (*Gorilla gorilla diehli, Gorilla beringei graueri, Gorilla beringei beringei*, and *Gorilla gorilla gorilla*), orangutan (*Pongo pygmaeus, Pongo abelii, Pongo pygmaeus*, and *Pongo abelii*), tiger (*Panthera tigris jacksoni, Panthera tigris amoyensis*, and *Panthera tigris sumatrae*), and hawksbill turtle (*Eretmochelys imbricata*).[1] Instead of being a spectator or consumer of these animals, the reader is compelled to confront them eye to eye—an equal and often unsettling exchange.

In April 2016, I was part of a collaborative residency that toured nine communities in Southeast Alaska to investigate climate change in the area while fostering dialogue and exchange within the communities we visited.[2] The experience shattered my egocentric view of a harmonious relationship with nature. While the developed world revolves around human greed and tries to make green improvements to our problematic industrialized world, indigenous philosophies require little modification. Humbled and awed, I learned about the delicate and exemplary way that these indigenous people weave themselves into the web of nature. Here the golden rules that operate within industrialized worlds are awkward, out of place.

1. https://www.worldwildlife.org/.
2. https://iialaska.org/tidelines/tidelines-2016/.

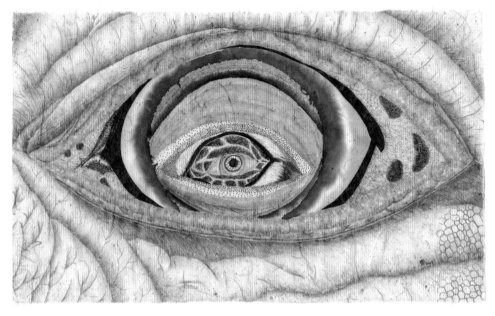

PLATE 7. Michelle Kuen Suet Fung, *Nahual*

For instance, in Alaska a bowl of salad can contribute more to global warming than red meat, after taking into account the fossil fuels used for transportation. This same feeling of displacement crept up as I read Carolyn Fornoff's powerful essay.

Most humans alive today are distant from nonhuman animals, except for domesticated pets and dead flesh packaged as calories. Nahual offers an alternative way to think of our relationship with animals, one that is less arrogant, less certain, and more humble.

I wanted to present Pachamama as a divine being in a moment of giving. With eyes closed, unconcerned with progress, advancement, or other human conquests, she is above those aims, one with the landscape that she embodies and manifests. The ring that surrounds her symbolizes the praise and attention that she receives as well as the voices of many who honor and worship her. Full of spring colors and a bright palette, she is surrounded by an abundance of earthly beauty.

< PLATE 8. Yellena James, *Pachamama*

Natasha Bowdoin

Plant Time

The natural world isn't always my overt subject, but its principles and spirit govern the way in which my work develops. Starting with an excerpt of found text, a totem, a gesture, or all three at once, my drawings grow instinctively, sprawling across a wall or tightening into a thicket of words and imagery. My cut paper collages and installations examine the potential intersections of the visual and literary, treating language and the biological as kindred phenomena. Drawing directly from the natural world or artists' renditions of it, my work marries a diversity of sources, including nineteenth-century botanical illustrations, 1970s patio prints, lunar surface maps, transcendentalist texts, early cartoons, and nineteenth-century adventure narratives. Past literary sources have included Lewis Carroll's *Alice's Adventures in Wonderland* and *Through the Looking-Glass*, poems by Jorge Luis Borges, and Walt Whitman's *Leaves of Grass*. Often incorporating layered and interwoven transcriptions, I merge found texts with drawn pattern, dissecting, fracturing, and reconfiguring these iconic texts into loose, free-flowing rhythms. This process endeavors to conjure the organic and the unruly while presenting an experience of the written word that acts more like fluid organism than fixed structure.

In my drawing *Plant Time*, I envision a landscape that is forever changing, subsuming humanity into the world around us. We dissolve within a meadow, forest, swamp, or desert, acquainting ourselves with our plant counterparts in a more deeply felt, sensory way.

< PLATE 9. Natasha Bowdoin, *Plant Time*

동서를 막론하고 가장 현명한 철학자는 "모든 것은 움직인다." 라고 말했습니다. 사물과 현상을 대상으로, 개념으로 고정해 버리면 경계 없이 유동적으로 늘 변화하는 현상에 대처하기 힘들고 또 오류를 범하기 쉽기 때문일 것입니다. 노자가 말한 '도 道'나 붓다가 말한 '공空' 역시 마찬가지로, 바로 그런 '늘 변화하는 세계', '모든 것이 가능한 세계'에 대한 진리를 말하고 있다고 생각합니다. 우리가 어떤 변화를 이루고자 할 때 역시 마찬가지로, '행복', '평화', '정화' 등의 개념을 고정하고 그것에 도달하기 위해 애쓸 것이 아니라, 행복한 행동 자체가 지금 당장 행복을 불러온다는 믿음을 갖고 행동해야 한다고 생각합니 다. 끊임없이 변화하려는 그런 노력, '기氣'의 성질을 그림에 담아보았습니다.

In both East and West, the wisest philosophers have said, "All things move." It may be because if we fix objects and phenomena as static notions, we cannot readily cope with the fluid and dynamically changing world. In the same way, Laozi's Tao and Buddha's Emptiness speak the truth about the "ever-changing world" and "the world where all things are possible." As such, I believe that when we look to change our lives, we must not pursue ideas such as "happiness" and "peace" as static goals but rather act with the faith that a happy disposition brings happiness to the present moment. I attempted to capture in this piece the nature of qi, which is the constant effort to change.

TRANSLATED BY SOO GO

PLATE 10. Moonassi, *Water-Wind (Qi)* >

The painting *Sehnsucht, in the Midst* depicts the moment of contradictory feelings and potential outcomes before one takes action. The figure is caught in between, grasping the arid rocks, looking toward a thriving green world, despite the psychological hold the central black void has on the figure as she seems to merge into it.

Sehnsucht embodies this feeling for me. With the destruction of our natural world, it is easy to feel despair. Yet at times we find the courage to remain keenly awake to this dark reality and simultaneously dream of a better coexistence with nature. Individual visions of our destination are personal, but their presence in our collective minds propels us forcefully toward a new world.

< PLATE 11. Nikki Lindt, *Sehnsucht*

Solastalgia

Kate Shaw

Sitting in a cafe in Ubud, Bali, I'm overwhelmed by traffic fumes. Each time I have visited, I have noticed the increase in traffic—especially SUVs clogging the streets. Heading out through the rice paddies, I see more and more hotels and restaurants catering to the ever-increasing tourist population. I commented on these changes to an Australian friend who has lived here for years, and she remarked, "I know, I'm so disappointed—imagine how the Balinese feel!" While my friend and I feel disappointed by these environmental changes, I wonder how the Balinese feel, to grow up in this area and witness such dramatic change. I walk past an old Balinese man, sheltered from the rain with the frangipani he is selling for the daily Hindu offerings. He gazes at the parade of cars and motorbikes, as though he is looking beyond them into another time in this same place, as though he is homesick for the Ubud that once was.

My work imagines solastalgia as a memory—perhaps embedded in human DNA—and depicts a cave in which "we" see through the eyes of an ancestor peering out from their shelter onto a landscape we have an uncanny longing for, like a primal homesickness.

PLATE 12. Kate Shaw, *Solastalgia* >

Sueño

Susa Monteiro

Ilustrando a palavra para um futuro ecotopico sueño, tentei encontrar um equilíbrio entre o homem e a sua relação com o espaço/mundo que o rodeia. O personagem que desenhei está numa espécie de limbo.

O homem que existe na luz e na noite. A palavra sueño, mais do que descrever uma acção, deveria ser uma forma de vida. Porque todos somos a coisa e o seu reflexo, o dia e a noite, o feminino e o masculino, e o lucido e o sonho. Todos somos nós mesmos e o meio que nos rodeia, indissociáveis.

Os homens estão agora divididos entre aqueles que respeitam e cuidam do meio que os rodeia, e aqueles que são movidos unicamente pelos interesses capitalistas e materialistas e cuja atitude perante a ideia de um pensamento ecológico é, a de fanáticos, que rejeitam as provas dadas a cada dia pela ciência.

Usei as cores do ocaso, o rosa e o azul cobalto, que representam aqui o próprio sueño, a esperança, e o preto na figura do homem representando a ideia de inexistência e tempo perdido.

Illustrating the ecotopian word sueño, I tried to strike a balance between man and his relation to the space/world that surrounds him. The character I drew is in a kind of limbo.

Humans exist in the light and in the night. Rather than describing an action, the word sueño should be a way of life because we are all both the thing and its reflection, the day and the night, the feminine and the masculine, the lucid and the dream. We are all both ourselves and the environment round us, inseparable.

We are now divided between those who respect and care for the environment around them, and those who are driven solely by capitalist and materialistic interests, whose attitude toward the idea of ecological thinking is that of fanatics who reject the evidence given to us every day by science.

I used the colors of the sunset, pink and cobalt blue, which represent here the dream itself and its hope, as well as black in the figure, representing the ideas of nonexistence and lost time.

PLATE 13. Susa Monteiro, *Sueño*

PLATE 14. Maryanto, *Terragouge*

Terragouge

Maryanto

I make paintings, installations, and murals of physical spaces and landscapes that are structured by social and political practices. My research traces how these places connect to history, ideology, and cultural identity.

This painting depicts a landscape in transformation, colonized by humans through the forces of industry. There are parts of Indonesia and Papua New Guinea where this scene is quite common because of the coal or gold mining industries.

In the age we live in, people have shaped the landscape according to their own needs, and the landscape is now constantly changing because of the desires and actions of humans. Not only is the land colonized but also capitalized and made profitable; it takes on monetary and financial value to corporations. This is what terragouge means to me.

Pronunciation: mish-nyuhkh (ɱiʃ:niəχ)

Part of Speech: Noun

Provenance: Gaeilge

Example: It took a bit of misneach to say no, but it helped with staying put, with humbly and ambitiously weathering the Anthropocene.

Misneach in Irish (Gaeilge) is most commonly translated into English as "courage." But the word belongs to an oral tradition and can mean much more in spoken conversations, where it encompasses a blend of courage, hopefulness, bravery, and spirit. It can allude to pushing forward, one paw in front of another, through doubt or snow. A bit brighter and less bellicose than bravery, misneach lingers longer with uncertainty, even as it carries on lifting each tired limb. Misneach knows tears and is changed by them, damp for trying. Some say that the sound of the word when uttered aloud can positively affect both the speaker and the listener: whispering "misneach" is like casting a spell.

The word misneach was put to work for nationalism in twentieth-century Ireland. During the war for Irish independence in the 1920s, a magazine named *Misneach* called for great works of world literature—especially those written in English—to be translated into Irish and made widely available to the public. Cultural recuperation and revival, the argument went, would require easy access to the canon as it had evolved during the centuries of British occupation. To dissuade people from reading English texts in a newly independent Ireland, the revivalists argued that high-quality translations must be readily available *as gaeilge*. The thrust of the proposition was that if Brontë, Dickens, Wordsworth, and others were offered in Irish, then fewer people would resort to the English versions to access these deemed-important books, fostering an informed literacy and national culture that did not rely on the English language.[1]

Misneach the magazine aimed to cultivate the thriving Irish-speaking community that nationalists dreamed of—a community that had endured in patches under the British and that might be now possible with a contested independence on the horizon. In seeking support for a celebratory revival of the Irish language, *Misneach* hoped to energize a postcolonial nation that would perform a precolonial, imagined sense of Irishness. It sought a new set of cultural practices that might support one another, enacting a present that might have existed under different, imaginary circumstances. Perhaps there was naivete in the optimism of the message—and such a culture never came to pass in the way it was envisioned—but there was misneach in the way they vividly conceived of an alternative contemporary moment and took steps to realize it.

Faced with multispecies extinctions amidst a changing climate, a bit of misneach might help us to imagine compelling possibilities for new presents. We might move to try and make a something else, even when there isn't enough fresh groundwater to reverse the losses and fill up the years of absence.

At the end of the twentieth century, my own sense of misneach was emerging at school. A friend and I were caught speaking English at our *gaelscoil*—the name for a school taught "through the medium of Irish." Our creatively conceived punishment was to spend a Friday afternoon translating passages from Shakespeare's *Macbeth* into Irish. As I remember, we loved speaking Irish to one another, and our occasional English chats accompanied our half-hearted (always failed) hopes to seem more rebellious than we were by breaking school rules. Our Friday afternoon disciplinarian instructed us to retain the iambic pentameter in Shakespeare's verse structure in our translations. I didn't know then that her task echoed *Misneach*'s 1920s mission to see Anglophone literatures available in Irish.

I do remember, though, starting with a speech of Lady Macbeth's. "Ach gabhail misneach agus fan anseo" was my poor but interested attempt at translating "But screw your courage to the sticking-place," Lady M's attempts to persuade her husband to murder the king.[2] My Irish version of her lines directly translates as "take misneach and wait here." Retrospectively, I don't think that she would have sounded convincing at all speaking my translation, and I doubt that Macbeth would have held his nerve without the sticking place, especially when it came to regicide. But Lady Macbeth probably wouldn't have fancied my kind of misneach anyway; there was too much vaulting ambition in the courage she sought. The misneach I'm proposing would have been happier by a window, soaking up the fair and foulness of a day and focusing on what the owl and the cricket had to say. To my knowledge, the complete works of Shakespeare are still not available in Irish, but I like to imagine that the combined efforts of decades of castigated students could be assembled—from fragments of penitence floating around recycling centers on scraps of faded copybooks—into one multiauthored text.

Since the magazine and the war for independence, the word misneach has subsequently been used to name organizations dedicated to celebrating and promoting the use of the Irish language, both in Ireland and among the Irish diaspora abroad. Today an activist group called Misneach na hÉireann remains committed to cultural revitalization, arguing for language rights as human rights, and shifting the earlier emphasis from making literary works available in Irish to making the language audible and visible in the present.[3] In 2015, for instance, Misneach na hÉireann campaigned successfully for Irish names to be permitted on Facebook, where previously the social media platform had sought proof that they were real.[4] In 2016, the group claimed to have defaced the image of Irish politician John Redmond on a Dublin city council poster that had been erected to commemorate the centenary of the 1916 Easter Rising, the nationalist rebellion against the British government in Ireland that led to the war for independence. According to Misneach na hÉireann, Redmond was responsible for the deaths of thousands during World War I because he had condemned the Rising and encouraged Irish citizens to fight with the British, hoping to strengthen Ireland's case for home rule. Members of Misneach na hÉireann spray-painted the number "35,000" in red graffiti over Redmond's face on the banner at the Bank of Ireland building at College Green to represent the estimated number of Irish people killed abroad, challenging the city council's reductive and revisionist history.[5]

Redmonds of many kinds watch us from commemorative banners in important places until someone with a bit of misneach and a can of spray paint cares enough to interrupt. How much misneach does it take to press down on the nozzle on a can of paint? Disruption may be at the heart of contemporary intersectional environmental politics, and it is worth remembering that it can take misneach in the shift from caring to acting, when actions come at a cost and outcomes are far from assured.

Misneach does not have a stiff upper lip; its heart quiverflaps on a windy day. It differs from the instructiveness of suck-it-up or get-on-with-it approaches to making it through bad weather. There is misneach living in our scrappy efforts, encouraging us where courage is not felt: if it's cold, put on some misneach (if the world has given you a coat). Misneach is not the same as a deep breath and a best foot forward, maybe because it doesn't know which foot is best.

In another part of Dublin today, there is a larger-than-life bronze sculpture called *Misneach* on a marble podium depicting a teenage girl riding a horse bareback. John Byrne's artwork was unveiled in 2009 and now poses with one hoof raised outside the local Trinity comprehensive school in Ballymun. In an interview, Bryne described *Misneach* as a "monumental celebration of youth," adding that the sculpture was about acknowledging "the hero in everybody, the hero in the normal and the ordinary and it's about courage."[6] Byrne said that he wanted to show that a girl from inner-city Dublin could be "as much a hero as a military hero."[7] He also wanted to celebrate the community tradition of riding horses bareback while referencing and subverting the ubiquity of equestrian sculptures found in town centers marking places and events of significance. So often it is military sculptures that mark places of importance, glorifying men of dubious repute on horses. Byrne wanted to show that a girl wearing a tracksuit had the potential to take the place of any one of these figures.

Bryne's *Misneach* also makes reference to Ballymun's strong historical links with horses, which have long been company for humans in this once suburban, now urban landscape. The local children's pastime and passion for horse trading and riding emerged in the 1980s, extending the legacy of the traveling community and traditional roles for horses in delivering fuel and milk to homes and businesses.[8] Some readers may remember the scene in Alan Parker's film *The Commitments* where a horse is transported up a tower block of flats in an elevator, making visible this continued tradition in popular culture.[9] After restrictive legislation via the Control of Horses Act in 1996, which affected concern for the welfare of horses and animal rights but in reality made the hobby prohibitive for young people who were already overlooked by the local authorities, journalist Fintan O'Toole described Ballymun's young "urban cowboys" as "a subculture struggling against extinction."[10] Tensions between human and animal health, deprivation, joy, and the "elemental freedom" felt by young people on horseback, who often described themselves as having a choice between horses and drugs, have characterized the ongoing debates about horses in Dublin's suburbs.[11] This is not to romanticize urban horse culture, which sometimes disregards animal suffering, but to think about the different interventions that might improve conditions for both people and animals living together. If the young people of Ballymun would have benefited from more amenities and assistance with the care of their horses, then the one thing they didn't need was to be stripped of their animal companions. It was to these debates, to the traditional culture of horse riding, and to local young people—young women in particular— attesting to their care for animals and potential for world making that Byrne was speaking when he made his statue for Ballymun and called it *Misneach*.

Donna Haraway recommends that one strategy for living in our ecoprecarious times might be to "stay with the trouble." She advocates for conscious and committed, responsible and responsive living in the tangle of the muddle that is now, rejecting both terrified or hubristic preoccupations with future apocalypse or salvation and smug or nostalgic backward glances toward awful or Edenic pasts.[12] Wade carefully into the depths of around, cognizant of forward and back, Haraway seems to say. Rereading the pages of Haraway's book waters my misneach. She reminds us that "it matters what matters we use to think other matters with," that "it matters what stories we tell to tell other stories with," that "it matters what thoughts think thoughts."[13] In gray I try to think with misneach.

Byrne's *Misneach* in Ballymun was cast from the leftovers of a sculpture of a British imperial war hero, Lord Gough, on a horse that used to stand in Dublin's Phoenix Park. The Gough memorial was unveiled in the park in 1878 to an audience that included a young Winston Churchill. In the original, Gough and his horse were cast in metal from cannons and guns that had been melted down and upcycled into art. Gough had been born in County Limerick, which is why the sculpture was erected in Ireland and not the United Kingdom, but militant Irish Republicans were unhappy with having a monument to an imperial war hero—especially one with as fierce and terrible a reputation as Gough—erected in Dublin, and they made multiple attempts to destroy the sculpture in the years that followed. Gough was beheaded and desworded in 1944 (the head was found in the River Liffey and reaffixed); the horse's right hind leg was similarly amputated in 1956; finally the memorial was completely destroyed and removed from the park in 1957. In 1990 its remains were sent to Chillingham Castle in Northumberland, England, where they were restored. When Byrne came to make *Misneach* for Ballymun, he was able to create a replica of the horse from the repaired Gough sculpture at Chillingham Castle and replace its rider. Not without some irony, Byrne's *Misneach* was completed in England and returned to Ireland by boat.

The horse that Toni Marie Shields—the seventeen-year-old chosen to model for the sculpture—rides in Ballymun today is an exact replica of the horse that once stood in Phoenix Park. The latest 3-D scanning technology was utilized to make an initial wax cast that caught the details of her clothing, Velcro runners, and long ponytail, which were more difficult to replicate than the textures usually found on bronze equestrian monuments. In Ballymun, a politician worried that *Misneach* was demeaning to the local people because it represented a girl in a tracksuit on a horse, an image he thought that the regenerated suburb was trying to shed. Shields, however, is reported as having said that the sculpture was for all of the young people of the place.[14]

Beyond attempting to elevate a local girl to the equivalent status of a military hero, Byrne's *Misneach* offers an ecological context for the horse that remembers the traditions and tensions highlighted by bareback horse riding in Ballymun. *Misneach* actually wrests the notion of equestrian sculptures from their military context, intentionally or not.

Shields diverts the horse from its route to the battlefield, modeling an alternative form of interspecies companionship. There is misneach in the hope of capturing the fabric of a tracksuit, in the serious work of fine-tuning the texture of a ponytail for posterity, in the heart of a woman moving toward a somewhere that isn't a war. Shields and the horse together demonstrate possibilities for living together in the trouble. They mark something of the persistence of disappearing traditions that know animals as companions in world weathering.

The original plan was that *Misneach* be moved to the town center of Ballymun, after plans for a new Metro North tramway were completed. The abrupt end to Ireland's Celtic Tiger affluence meant that the Metro North never happened, though; as of 2017, *Misneach* was still outside of the Trinity comprehensive secondary school, there to witness students passing on their way in and out of school: learning words ("say 'MISH-nock.' It's like 'courage' in English"), animals ("when we take you to visit the Natural History Museum, you'll see skeletons of the *Megaloceros giganteus*, giant Irish deer that once roamed the country when the land was still connected to mainland Britain and Europe. They are thousands of years extinct"), and histories ("for hundreds of years, Ireland was under English rule, but before this and after the deer roamed the land, people spoke Irish all the time"). *Misneach* is the feeling of walking into a new school for the first time. The word misneach is carved into the stone plinth that holds Shields and the horse, naming the idea it hopes to disperse.

Here misneach is a loanword that can't really be on loan because it was always already around in the leftovers of a living language, not quite forgotten. But even if the word is indigenous to Ireland, the language usually takes some effort to learn. And although Trinity's students, like many others in Ireland today, speak mostly English to one another and learn their lessons through English at school, the Irish language retains a prominent place on a curriculum that begins with *Gaeilge, Béarla,*

Mata—Irish, English, and maths. When I try to think about my own learning at school, misneach floats somewhere in the dreamy repetition of past learning. These days my misneach bubbles at the edges of what it means to live abroad, not to speak with others but to recall words that help with unbearable times and things (Ireland's last freshwater pearl mussel buried itself in the substrata of a polluted river on its way to extinction—is saying it, writing it, grieving it enough?). Misneach is what I have loaned myself from the language, surfacing when I need it most and expect it least. It's a secret remembered quietly in company, breathing in the trouble and filling me with something like courage that isn't quite courage to help with hopeful flailing, staying—things I thought I couldn't do.

Maybe if—or maybe because—the stories that we tell our stories with matter, my misneach remembers with a magazine, Lady Macbeth, red spray paint, an artist, urban horses, a young woman, gunmetal and bronze, words from books and conversations.[15] And somewhere a pigeon shits on a statue. Somewhere buddleia thickens out the cracks. A sycamore prongs through aging tarmac. A horse gets away. Take heart. Take misneach.

In this razor-wire world of terrifying weather forecasts, the amalgam of wary courage, hope, and spirit that misneach encompasses strikes me as very much needed and on offer to those who don't feel brave.[16] Misneach might mean an ethical, affective impulse that follows but remembers the moment before an ecoprecarious inbreath. It might mean the impetus for forward movement—whether the wearer believes in progress or not. Misneach might be the nudge you need to keep going on uncertain but certainly stormier days. I'm scared. I'll loan you mine.

Rén
ANOTHER PATH

NOTES

1. Máirtín Mac Niocláis, *Seán Ó Ruadháin: Saol agus Saothar* (An Clóchomhar: Baile Átha Cliath, 1991), 108–9.

2. William Shakespeare, *Macbeth* (1606), ed. Sandra Clark and Pamela Mason (London: Arden Third Series, Bloomsbury, 2015).

3. For more on this provocation, see Niamh Nic Shuibhne, *Cearta Teanga mar Fíorchearta Daonna? Language Rights as Human Rights?* (Dublin: Bord na Gaeilge, 1999).

4. "Misneach: Success over Facebook Campaign," *Celtic League*, September 30, 2015, https://www.celticleague.net/.

5. "Activists Draw Over John Redmond's Face on Controversial College Green Banner," *Journal*, March 25, 2016, https://www.thejournal.ie/.

6. "Controversial Statue Unveiled in Ballymun 2010," RTÉ (Raidió Teilifís Éireann), September 17, 2010, https://www.rte.ie/.

7. "Misneach: A Monumental Celebration of Youth," PublicArt.ie, n.d., https://publicart.ie/.

8. For an expansive discussion of Ballymun's horse culture, see Lynn Connolly, *The Mun: Growing Up in Ballymun* (Dublin: Gill & Macmillan, 2006), 181–99.

9. *The Commitments*, dir. Alan Parker (20th Century Fox, 1991).

10. Fintan O'Toole, in Perry Ogden, *Pony Kids* (London: Jonathan Cape, 1999).

11. See Finton O'Toole, "Pony Kids, Urban Cowboys," *Independent*, February 6, 1999, https://www.independent.co.uk; *Pony Kids, Dublin's Urban Young Cowboys: A Documentary by Magali Chapelan*, https://vimeo.com/67776714; "Ireland's Homeless Horses Face Mass Cull," *Guardian*, October 15, 2010, https://www.theguardian.com/.

12. Donna J. Haraway, *Staying with the Trouble: Making Kin in the Chthulucene* (Durham, N.C.: Duke University Press, 2016), 1.

13. Haraway, *Staying*, 12.

14. Jason O'Brien, "Sculpture of Tracksuit-Clad Rider Is Mane Attraction," *Irish Independent*, September 18, 2010, https://www.independent.ie/.

15. Haraway, *Staying*, 12.

16. I'm thinking of weather forecasts here with Mike Hulme's idea of a world where the notion of stable climates will no longer be accessible to us. Hulme, *Weathered: Cultures of Climate* (London: Sage, 2017). Also, Astrida Neimanis and Rachel Loewen-Walker propose a feminist new materialist take on weathering as a way to encounter climate in "Weathering: Climate Change and the 'Thick Time' of Transcorporeality," *Hypatia* 29 (2014): 558–75.

Pronunciation: na-wal (naːwɔːl)

Part of Speech: Noun

Provenance: Mesoamerican worldviews

Example: My nahual peered briefly through my eyes today.

Along with rising sea levels and extreme weather events, one of the biggest crises accompanying the Anthropocene is the advent of a sixth mass extinction. Species are disappearing at an unusually rapid clip: conservative estimates predict that one-third of all species will go extinct over the next hundred years.[1] Even many of those that do not face imminent extinction have been experiencing rapid population decline. This defaunation adversely affects broader ecological networks: fewer animals to pollinate and disperse seeds contributes to the loss of plant life.[2] How did we get here? Researchers point to proximate causes such as pollution, climate change, and habitat fragmentation, all of which are propelled by human activities and the organizing logic of capitalism, which requires Cheap Nature to drive the accumulation of wealth.[3]

Underlying such practices is an assumption central to Western humanist thought: humans and nonhumans occupy separate, discrete realms of activity and knowledge. Since Aristotle, philosophers have endeavored to identify the key characteristics that set humans apart from other animals. The human abilities to speak, reason, create art, and feel shame or boredom have been put forth as discriminating qualities that elevate our species above others. These qualities uniquely endow humans with personhood—with individuality as well as political and ethical status.[4] In this way, life is divvied up into an easily graspable hierarchy, which allows us to value and protect human life while deeming animal life anonymous and expendable.

Yet the onset of the sixth mass extinction prompts us to rethink these basic premises about who and what should be valued. It urges us to look for other models that conceptualize the relationship between human and nonhuman not as disconnected but as intimate and enmeshed. One such model can be found in the Indigenous practice of nahualism in Mexico and Central America. Dating back to pre-Columbian times, nahualism asserts that each human is born linked to an animal alter ego, her coessence or nahual (alternatively, *nagual* or *nawal*). The nahual accompanies that human over the course of her entire life. The human and animal pair shares a soul or consciousness; they have the same breath but adopt different bodily forms.[5] Nahualism demonstrates an approach to ontology—the nature of being—that dramatically diverges from Western models. It formulates human and nonhuman life as inextricably intertwined. This connection is not just external (the connection we might feel when we see or encounter "nature"); it is also innate, contained within the very self. An individual life cannot therefore be understood to be bounded or autonomous because it is already a multiplicity; each self is not either human or nonhuman, but both.

The Maya K'iche' leader and Nobel Prize winner Rigoberta Menchú dedicates an entire chapter of her testimonial text, *I, Rigoberta Menchú*, to the nahual.[6] She explains that this belief is vital to K'iche' identity:

> Every child is born with a *nahual*. The *nahual* is like a shadow, his protective spirit who will go through life with him. The *nahual* is the representative of the earth, the animal world, the sun and water, and in this way the child communicates with nature. The *nahual* is our double, something very

important to us. We conjure up an image of what our *nahual* is like. It is usually an animal. The child is taught that if he kills an animal, that animal's human double will be very angry with him because he is killing his *nahual*. Every animal has its human counterpart and if you hurt him, you hurt the animal too.[7]

Because numerous Indigenous groups throughout Mesoamerica practice nahualism, its details vary accordingly. Menchú explains that in the K'iche' tradition, the calendar determines one's nahual.[8] Every day of the month corresponds with a specific wild animal, plant, or natural phenomena: jaguar, eagle, maize, wind, even lightening. The entity that correlates with a person's birth date becomes his or her lifelong companion. They are the same spirit but inhabit different bodies. That is, the soul occupies two different material forms, one human and one nonhuman. These two bodies share the same essence, but there is no communication between them. They cannot access each other's perspective without the help of a shaman (through the ritualistic consumption of plants like peyote), through dreams,[9] or at the moment of death when they can briefly see through the nahual's eyes. Because this pair is the same being, the term nahual doesn't just signify the nonhuman alter but also the human herself. The nahual describes a unity of being that is also a duality, a self that is also an other.

The relationship with the nahual is understood in various ways, depending on the group or person describing it. It is one's "shadow," "double," or "protective spirit,"[10] one's defender or vigilant guard,[11] a coessence,[12] or an "other I" that is "molten into the human being, perched onto his back."[13] The coastal Mixtec people describe those who possess a nahual or coessence as *uvira*, which translates as "two-man" and evokes "an abstract image of plural singularity or unity."[14] Zapotec poet Victor Terán directly links the nahual with the sensation of being in love, describing it as an intimacy that is invisible yet palpable: "It will feel like ants tickling at your soul, / but you will never see it."[15] In Terán's poem, *nagual* is a translation of the Zapotec word *xquenda*. The translators note that another possible translation might be "one's characteristic."[16] The nahual is a person's constitutive trait, a foundational element of identity and self.

To nonindigenous peoples, the idea that our self does not belong solely to us but also exists in another body—not to mention the body of a different species—is hard to grasp. This is particularly the case because in the Western philosophical tradition, the self is assumed to be a unified, contained being that is fundamentally separate from other beings. Nahualism scrambles this logic and undermines the supposed autonomy of the self by suggesting that a single self or soul can take on various material forms. To further clarify this concept, it is helpful to look to the novel *Men of Maize* (1949) by the Nobel Prize–winning Guatemalan author Miguel Ángel Asturias. While Asturias was not himself Indigenous, his novel draws from foundational texts of Mesoamerican mythology including the Popol Vuh and the Chilam Balam.[17] In one chapter, the character Gaudencio Tecún explains the parallel deaths of a deer and a local curer (a shaman, or spiritual guide and healer) to his brother Uperto:

"The curer and the deer, for your information, were one and the same person. I fired at the deer and did in the curer, because they were one and the same, identical."

"I don't get it. See if you can explain it to me. The curer and the deer," Uperto raised his hand and put his two middle fingers together, "they were like seeing one fat finger made up of two fingers."

"No, not that. They were the same finger. Not two, one. The curer and the deer were like you and your shadow, you and your soul, you and your life breath. . . . Like two drops of water in one swig."[18]

The realization that man and deer are one and the same perplexes Uperto. As he grapples with the implications of this insight, he lyrically articulates the nahual's multiplicity of experience:

> To think that the deer and the curer were one single being
> was so difficult for him that at times he held his head, fearful
> that his own common sense might be turned. That dead body
> had been a deer, and the Deer of the Seven-Fires had been a
> man. As a deer he had loved does and had had fawns, baby
> deer. His male nostrils [had been] in the algebra of stars, the
> bluish coats of the does, soft fur toasted like the summer,
> nervous, shy, susceptible only to fugitive loves. And as a man,
> when he was young, he had loved and pursued females, he'd
> had little human children full of laughter, whose only defense
> was their weeping. Which had he loved more, the does or
> the women?[19]

The curer and the deer outwardly manifest divergent material forms and experiences. Yet the apparent differences between these two bodies hide the nahual's ontological undifferentiation. There is spiritual unity with bodily diversity. The deer is not "just an animal, but an animal that was a person," a "deer who [is] not only deer."[20] This framing of man and deer as one and the same deconstructs human exceptionalism. Uperto's lingering question of "which he had loved more" leads the reader to realize that human love is no more real than cervine. Both experiences are equally consequential to the curer/deer's life.

What becomes clear after this passage in *Men of Maize* is that every nonhuman creature might also be human. Put another way, every animal, plant, or natural formation (like a river, mountain, or cave) exists in a latent state of possible personhood. Brazilian anthropologist Eduardo Viveiros de Castro has studied this facet of Amerindian cosmologies. He explains that Indigenous animist worldviews approach "every object [as] a subject in *potentia*."[21] Therefore, "every being encountered by a human over the course of . . . his or her own life may suddenly allow its 'other side' (a common idiom in indigenous cosmologies) to eclipse its usual non-human appearance, actualizing its latent humanoid condition."[22] For the K'iche', the nahual is that "other side," that humanoid soul that can dress or disguise itself in different external garbs. The external body does not define identity but is a removable cloak that conceals the unchanging essence underneath.[23]

If the West has traditionally separated animals from humans on the basis of what they supposedly lack—be it reason, soul, art, or language—Amerindian societies proceeded in the opposite direction. They propose that personhood, typically attributed exclusively to humans in the West, is a position that can be occupied by other species as well. If humans think about themselves as persons, in the Amerindian tradition, then so do animals. As Brazilian literary scholar Idelber Avelar puts it, "When a jaguar sees you, he is the one who is a person."[24] The animal assumes the position of the perceiving subject, and therefore its behavior is not dictated by instinct but shaped by cultural norms. Viveiros de Castro elaborates this point: "Being people in their own sphere, non-humans see things as 'people' do. But the things that they see are different: what to us is blood, is maize beer to the jaguar . . . what we see as a muddy waterhole, the tapirs see as a great ceremonial house."[25]

Nahualism is part of a broader worldview positing that humans and animals coexist within a shared ecology. Menchú explains, "Man is part of the natural world. There is not one world for man and one for animals, they are part of the same one and lead parallel lives."[26] This affirmation of human/nonhuman inseparability is, after all, not so distant from findings in modern science that affirm that over half of the human body is made up of nonhuman DNA: the latest estimates suggest that for every human cell, the human body contains 1.3 bacteria cells.[27] Such revelations mirror the belief in coessence: from life to death we coexist with other creatures, whether or not we are aware of their existence. Our self is also that of an other.

To return to Menchú's testimonial text, she chooses not to reveal what her nahual is to her readers. Her decision not to disclose this aspect of her identity tells us two things. First, it tells us that the nahual represents the most personal and sacred expression of one's self. K'iche' poet Humberto Ak'abal similarly cautions against sharing this most intimate of secrets: "To tell someone about one's *nahual* is to remove one's clothes and allow everyone to see one naked."[28] Second, this secrecy is central to Menchú's broader approach to explaining indigenous culture to Western audiences. To counteract exploitation and appropriation, she chooses to reveal certain aspects while keeping others confidential. Menchú states, "We often find it hard to talk about ourselves because we know we must hide so much in order to preserve our Indian culture and prevent it being taken away from us. So I can only tell you very general things about the *nahual*. I can't tell you what my *nahual* is because that is one of our secrets."[29] Because of the violent and intertwined histories of colonialism, neocolonialism, racism, and genocide in Guatemala, outsiders have proven themselves to be unreliable recipients of Indigenous knowledge. As such, silence is an important tool. Ak'abal's poem "Stones" beautifully condenses this sentiment: "It is not that stones are mute; / they keep silent."[30] In this interpretation, the absence of speech is not a privation but a plenitude. The nonhuman world speaks, but it chooses not to speak to us.

The intimacy between a person and her nonhuman alter also gives shape to a different mode of affectively and ethically relating to animals. Whereas in Western thought it is common to foster emotional and ethical attachments with domesticated animals, nahualism is not restricted to the pet economy. Menchú explains that in Mayan communities the nahual has a pedagogical function; it teaches kindness across species from an early age. The intimate identification with animals validates their status as subjects, and subsequently their ethical consideration. By understanding each human self (the spirit) as also existing in an animal or nonhuman form, nahualism redirects our ethical relationship with nature from one that is determined by an individual's moral code to an ethics that is based on relationality, structured around the recognition that the nonhuman also possesses personhood or a point of view.

Rather than shore up a notion of the sovereign human subject, the belief in the nahual formulates a vision of transsubjectivity. From birth, each individual is aware of a connectivity to nonhuman life. Everything is connected, even if that connection cannot be seen. This transsubjectivity is felt and imaginatively embodied. It is an alliance that is both material and abstract, that crosses but also maintains boundaries. This philosophy prefigures and anticipates the move made by contemporary posthumanist scholars such as Stacy Alaimo, Karen Barad, and Rosi Braidotti, who argue in favor of a more porous conception of the human subject, one in constant becoming with other, nonhuman forces.

It is important to note that historically nahualism has been twisted to fit racist interpretations of Indigenous peoples as more "natural" and "irrational" than their Western colonizers or ladino (nonindigenous) counterparts. Nahual is derived from the Nahuatl word *nahualli*, which roughly translates as "magician." Unlike Menchú's identification of the nahual as a universal trait, in this etymology, it is the magician or shaman who has the privilege of shape-shifting between human and animal form. In colonial sources, the shaman is negatively construed as a "witch" or "wizard" who is therianthropic (part human and part animal).[31] In spite of colonists' attempts to eradicate non-Christian practices, these spiritual guides used the peyote plant to allow others to momentarily see through their nahual's eyes. The consumption of the hallucinogenic plant during spiritual ceremonies "was regarded as a method of throwing the individual out of himself and into relation with the supernatural."[32]

The fact that nahualism continued in spite of the colonial introduction of Catholicism led it to be perceived as a transgressive and even diabolical practice. Many sources written in English double down on the assertion that nahualism is a dark art. Scottish folklorist and occult scholar Lewis Spence explained in his 1930 book *The Magic and Mysteries of Mexico* that after a baby was baptized by the Catholic Church, some would take their infants to a shaman, "who, by some sleight of devildom, nullifies the sacred power of the holy water and bestows on the child a nagual or beast-guardian, a spiritual guide and mentor in animal shape."[33] The association of nahualism with dark sorcery or "devildom" illustrates how alternatives to Western modes of thought have been disparaged, censored, or repressed through violent and discursive means.

Menchú's insistence on maintaining the secrecy of her nahual adds an important note of caution as we consider nahual as a loanword for the Anthropocene. Her silence is a form of resistance that indexes the continuing struggles of Indigenous peoples, which is as relevant today as ever. In addition to facing racial discrimination and high rates of poverty, Indigenous environmental activists continue to be murdered at alarming rates throughout Latin America. Honduras is one such example; from 2010 to 2016, over 120 activists were killed for resisting the efforts of corporate and state interests to meet the global demand for palm oil, timber, and minerals.[34]

To think with Indigenous cosmologies as nonindigenous peoples is to challenge ourselves to be like Uperto, the character in Asturias's novel who holds his head in bewilderment as he learns about the nahual. It is to push ourselves to stretch our minds beyond anthropocentric paradigms of personhood, such that we become uncomfortable or even lose our common sense. To think with the nahual is to take the plunge of considering that the body—whether human or animal—might not determine identity but rather may be a removable garment or piece of clothing. It is to think about the coexistence of unity within difference (one self/consciousness inhabiting many bodies) and conversely of difference within the seemingly singular self (one body containing multiple selves).

Stacy Alaimo notes that in Western thought, neither ethics nor politics has "allowed space for concern over nonhuman lives."[35] Nahualism offers an important counterpoint, a way of conceiving of the human self as inherently connected to nonhuman life. Nahualism coincides with contemporary animal studies' interest in extending personhood to nonhuman animals and problematizing the dualistic conception of nature and culture as distinct domains. Because Amerindian thought articulates a paradigm in which every nonhuman thing is seen as a human *in potentia*, it challenges Western solipsism and promotes a horizontal ethics. In this way of thinking, human and nonhuman lives carry the same worth. Indeed, they are entangled such that the death of the animal double means the death of oneself. As we look for sources that model nonhierarchical, nonexploitative ways of relating to the nonhuman world, nahualism offers a provocative point of departure. By imagining ourselves inside the body of an other, we can begin to feel the urgency necessary to respond to the ongoing sixth extinction.

NOTES

1. Philip Cafaro, "Three Ways to Think About the Sixth Mass Extinction," *Biological Conservation* 192 (2015): 387.

2. Gerardo Ceballos, Paul R. Ehrlich, and Rodolfo Dirzo, "Biological Annihilation via the Ongoing Mass Extinction Signaled by Vertebrate Population Losses and Declines," *Proceedings of the National Academy of Sciences of the United States of America* 114, no. 30 (2017): E6089–96.

3. See Jason Moore, *Capitalism in the Web of Life: Ecology and the Accumulation of Capital* (New York: Verso, 2015).

4. For more on the animal in Western philosophy, see Matthew Calarco, *Zoographies: The Question of the Animal from Heidegger to Derrida* (New York: Columbia University Press, 2008).

5. It is helpful to briefly clarify the difference between the nahual and the totem in Mesoamerican worldviews. Whereas a nahual is a living animal attached to a living human being, a totem, by contrast, is an animal that is recognized by a group to be a sacred ancestor or protective spirit. Claire Pailler, "Tótem o nahual: El bestiario

nicaragüense en la afirmación de una identidad nacional a partir del Movimiento de Vanguardia," *Centroamericana* 21 (2011): 55.

6. For her leadership in the decade-long campaign to end the ongoing genocide of Guatemala's indigenous population, Rigoberta Menchú was awarded the Nobel Peace Prize in 1992. Menchú's testimony, *I, Rigoberta Menchú*, became embroiled in controversy in the U.S. academy in the 1990s, primarily over the question of its veracity. For a comprehensive account of this debate, see Arturo Arias, ed., *The Rigoberta Menchú Controversy* (Minneapolis: University of Minnesota Press, 2001).

7. Rigoberta Menchú and Elisabeth Burgos-Debray, *I, Rigoberta Menchú*, trans. Ann Wright, 2nd ed. (New York: Verso, 2009), 18.

8. Other ethnographies offer different accounts of how the nahual is determined. For instance, Gonzalo Aguirre Beltrán describes a Nahua ritual in which a newborn is left overnight in a sacred site surrounded by ash. The following morning, a shaman examines which footprints have been stamped into the ash, and that creature becomes the infant's nahual or protective guide for the rest of his or her life. Idi M. Flores Durán, "Apropiación mágico-religiosa del medio ambiente natural. Los nahuales, sabios con el poder de transformarse en animal," *Gazeta de Antropología* 22 (2006): article 4, http://digibug.ugr.es/handle/10481/7084.

9. John Monaghan, "The Person, Destiny, and the Construction of Difference in Mesoamerica," *RES: Anthropology and Aesthetics* 33 (1998): 142.

10. Menchú and Burgos-Debray, *I, Rigoberta Menchú*, 18.

11. Mario Molina Cruz, *Xtille Zikw Belé Ihén bene nhálhje ke Yu' Bza'o/Pacho Culebro y los naguales de Tierra Azul* (Mexico City: Consejo Nacional para la Cultura y las Artes, 2007), 352.

12. The coinage of the term "coessence" to describe the nahual in English is attributed by John Monaghan to Ester Hermitte. Monaghan, "Person, Destiny," 141.

13. Pablo Antonio Cuadra, *El hombre, un Dios en el exilio* (Managua, Nicaragua: Fundación Internacional Rubén Darío, 1991), 1.

14. Monaghan, "Person, Destiny," 142–43.

15. Víctor Terán, "Six Variations on Love," in *Words of the True Peoples: Anthology of Contemporary Mexican Indigenous-Language Writers*, ed. Carlos Montemayor and Donald Frischmann (Austin: University of Texas Press, 2005), 2:65.

16. Terán, "Six Variations on Love," 65.

17. In spite of the fact that Asturias was awarded the Nobel Prize for literature in 1967, his work has had a fraught reception in Latin America. Several prominent Maya intellectuals, including the K'iche' poet Humberto Ak'abal, attacked Asturias for presuming to speak for the Maya people, as well as for the racism in his undergraduate thesis on the "Indian problem." Menchú, on the contrary, revindicated his work, praising its "respect for difference." Arturo Arias has written extensively on this debate, noting that ironically this rejection of Asturias validates essentialist notions of Indigenous identities. While some critiqued his literature for not being

authentically Indigenous, other Latin American intellectuals (e.g., Ángel Rama and Emir Rodriguez Monegal) rejected it for copying Indigenous sources too closely (i.e., as being too much like Indigenous sources). Arias argues that this uproar reveals Latin American cosmopolitan prejudices: racial bias against Indigenous subjectivity, geographical bias against Central America, and *machista* prejudice against queerness. Arturo Arias, "Constructing Ethnic Bodies and Identities in Miguel Angel Asturias and Rigoberta Menchú," *Postmodern Cultures* 17, no. 1 (2006), http://pmc.iath.virginia.edu/issue.906/17.1arias.html.

18. Miguel Ángel Asturias, *Men of Maize*, trans. Gerald Martin (Pittsburgh, Pa.: University of Pittsburgh Press, 1995), 56–57.

19. Asturias, *Men of Maize*, 63.

20. Asturias, *Men of Maize*, 68.

21. Eduardo Viveiros de Castro, "Immanence and Fear: Stranger-Events and Subject in Amazonia," trans. David Rogers, *HAU: Journal of Ethnographic Theory* 2, no. 1 (2012): 41.

22. Viveiros de Castro, "Immanence and Fear," 31.

23. Whereas Western culture perceives the unity of nature (universality of substance) and the multiplicity of culture, Amerindian multinaturalism inverts this equation, perceiving spiritual unity and corporeal diversity. For more on this, see Eduardo Viveiros de Castro, *Cosmological Perspectivism in Amazonia and Elsewhere: Four Lectures Given in the Department of Social Anthropology, University of Cambridge, February–March 1998* (Chicago: University of Chicago Press, 2012).

24. Idelber Avelar, "Amerindian Perspectivism and Non-human Rights," *Alter/nativas* 1 (2013): 13.

25. Eduardo Viveiros de Castro, "Cosmological Deixis and Amerindian Perspectivism," *Journal of the Royal Anthropological Institute* 4, no. 3 (1998): 477–78.

26. Menchú and Burgos-Debray, *I, Rigoberta Menchú*, 22.

27. Ron Sender, Shai Fuchs, and Ron Milo, "Revised Estimates for the Number of Human and Bacteria Cells in the Body," *PLoS Biology* 14, no. 8 (2016), https://doi.org/10.1371/journal.pbio.1002533.

28. Irene Piedra Santa, "Humberto Ak'abal, the Guatemalan Weaver of Words," *Skipping Stones* 19, no. 5 (2007): 33.

29. Menchú and Burgos-Debray, *I, Rigoberta Menchú*, 22.

30. Piedra Santa, "Humberto Ak'abal," 33.

31. Franciscan Spanish friar Bernardino de Sahagún's pioneering enthnographic text of Aztec beliefs, the *Florentine Codex*, notes, "The *naualli*, or magician, is he who frightens men and sucks the blood of children during the night. He is well skilled in the practice of this trade, he knows all the arts of sorcery (*nauallotl*) and employs them with cunning and ability; but for the benefit of men only, not for their injury" (book 10, chap. 9). Sahagún's interpretation leans in two opposing directions: the

naualli is wicked but also skilled and benevolent. Bernardino de Sahagún, *General History of the Things of New Spain: The Florentine Codex*, trans. Charles E. Dibble and Arthur Anderson (Santa Fe, N.M.: School of American Research, 1975).

32. Daniel G. Brinton, "Nagualism: A Study in Native American Folk-Lore and History," *Proceedings of the American Philosophical Society* 33, no. 144 (1894): 9. Shamanic rites related to nahualism are still practiced today by groups in Mexico and Guatemala, as can be seen in Nicolás Echeverría's excellent 2014 documentary *Eco de la montaña* (Echo of the mountain).

33. Lewis Spence, *The Magic and Mysteries of Mexico, or The Arcane Secrets and Occult Lore of the Ancient Mexicans and Maya* (London: Rider, 1930), 165.

34. "Honduras: The Deadliest Country in the World for Environmental Activism," Global Witness, January 31, 2017, https://www.globalwitness.org/.

35. Stacy Alaimo, *Exposed: Environmental Politics and Pleasures in Posthuman Times* (Minneapolis: University of Minnesota Press, 2016), 10.

Nakaiy

Kira Bre Clingen

Pronunciation: na-kai (nɛkaɪ)

Part of Speech: Noun

Provenance: Dhivehi, Maldivian fishermen

Example: When developing a nakaiy calendar, Marblehead residents outlined twelve weather periods corresponding to optimal northeastern planting schedules and planned festivals to celebrate the beginning of each microseason.

Perched on the bow of a bamboo dhoni, Mr. Ali Rishman, the director of one of the few environmental NGOs in the Maldives, extolls the rhythm of the monsoon as we cut through the open sea beyond the capital of Malé. Pradeep, a fisherman from neighboring Villimalé, grunts as he mans the tiller. The dhoni, the traditional fishing vessel of the islands, has no keel, leaving it largely at the mercy of the winds and the navigator's brute strength. He points to five wispy cumulus clouds lingering over the cerulean lagoon. "Leftovers from the nakaiy Furahalha," he explains. "Now it's Hiyaviha, the skies should be clear, the tuna catch should not be so good. The market is too full for this time of year."

Furahalha and Hiyaviha are two of the twenty-six nakaiy periods. Meaning "constellation" in Dhivehi, nakaiy are divisions of the Gregorian calendar by predictable weather patterns.[1] Located in the crosshairs of the Indian Ocean monsoons, the Maldivian climate generally obeys the outline of wet and dry seasons familiar to the tropics. Lasting thirteen to fourteen days, each nakaiy is given a name in Dhivehi, and each is characterized by its wave height, cloud pattern, tidal surge, current strength, and sea surface temperature, creating a reliable framework of expected microclimatic conditions.

The atolls were first explored and settled by sailors carried on dry-season winds from the Indian subcontinent nearly three thousand years ago; these same monsoon winds still regulate transportation and bring rain to the country.[2] Yet the winds are not always benevolent, and to this day most islands within the twenty-six atolls remain uninhabited, incapable of sustaining human life. Investing in permanent settlements beyond thatched bamboo huts is a risky undertaking for families who live at the mercy of the tides and gales. Flooding and inundation are commonplace across the nation, whose highest point stands just 2.4 meters above sea level, making the Maldives the lowest-lying country in the world.[3] Indeed, the vastness of the surrounding ocean dominates the lives of the 325,000 islanders. The white sand islands themselves, comprising just 1 percent of the total area of the Maldives, seem no more permanent than crashing whitecaps, ringed by turquoise waters that abruptly turn a deep ultramarine where the fringing reef gives way to nothing but the deep sea. This sense of physical fragility has been exacerbated by a series of climatic events over the last two decades, including storm surges, coastal flooding, and the depletion of the fresh-water lens that floats above sea level.[4] The frequency and intensity of these phenomena is likely to increase, with the Global Environmental Fund declaring that "no settlement on the Maldives is entirely safe from the predicted impacts of climate change."[5]

Only the capital, Malé, the sole urban island in the country, is guaranteed the delivery of fresh produce, imported consumer goods, and, most importantly, potable water. As such, one third of the population lives in Malé, which is one of the most densely populated cities in the world. Tenants routinely sleep in eight-hour shifts to afford the rent, which is disproportionately high compared to other islands. Residents of far-flung islands have been relocated to the capital atoll in an effort to concentrate the population where centralized services, especially fresh water, can be more easily provided.[6] The Maldivian government and people must balance the country's unparalleled ecological fragility with the demands of civil society, a negotiation that nakaiy have helped to facilitate.

Indigenous Maldivian fishermen have relied on nakaiy for generations. Along with fourteen thousand other fishermen, Pradeep uses nakaiy to predict and plan his fishing, transportation, planting, and social life.[7] Originally conceived by seamen observing their environment, the hyperlocalized microseasons have moved beyond the working vernacular of fishermen and are now embedded within the social structure of the atolls. As the terms expanded into the common lexicon, each microseason was attributed new dimensions of meaning, including superstition and ritual beyond the original abiotic descriptions. Today all Maldivians use nakaiy in conversation, regardless of their profession, gender, caste, or creed.

Nakaiy are synonymous with beginnings. In late February, sharp bolts of lightning are let loose during the microseason of Furabadhuruva; Maldivians bring the fire to earth, burning small plants and bushes to ready the soil for planting during Burunu, a nakaiy in late April.[8] During Reyva, the nakaiy commencing on March 26, potatoes sprout hairs that rise above the sandy soil, which are clipped and dried for use in stews eaten at festivals that celebrate the beginning of the southwest monsoon. Adha signals the advent of the fishing season on June 17, accompanied by the transplanting of yams and any other vegetables that have managed to take root in the nutrient-poor, sandy soil. Assidha brings the first storm of the wet season on April 8, when Maldivians step outside to welcome the rains as well as marriage proposals. Even Pradeep's son-in-law, an international businessman born and raised in Malé, adhered to the tradition and asked Pradeep's daughter to marry him on April 10. Pradeep jokes that all his grandchildren are likely to be born during Uturahalha, a nakaiy in late January, nine months after the engagements and hasty weddings.

Passed down through the oral tradition, nakaiy are a repository of observational citizen science. In their most abbreviated definitions, they relate abiotic factors to the natural resource economy, primarily fishing and subsistence agriculture. The fishermen who originally coined the terms weighed the optimal times to fish against the best weeks to come ashore to plant crops, attempting to provide year-round nutrition in a temperamental environment exacerbated by seasonal fish migration and substandard soils. Yet underlying the shortened description of each microseason is a deeply embedded knowledge of the way that weather patterns affect the ecology of the islands and surrounding oceans. Viha, beginning on November 1, ends the strong winds of Hey before it. Viha is a calm period during which albacore, known colloquially as "fish," spawn every two days. After Viha, the winds gradually increase in tempo until the arrival of the northeastern dry season monsoon on December 10, when Mula's fierce gusts push fleets of dhoni away from the southern atolls where albacore plankton are dispersed by the current and the catch is depleted.

Maldivians' categorization of natural phenomena into discrete periods represents an attempt to eradicate the fear of instability and violent natural chaos across a seascape that is largely beyond human control.[9] The act of assigning names to abiotic features is an attempt to grapple with the climate and increase the sense of human power over the environment. Nakaiy act as social guidelines that regulate behavior and promote human well-being. By fishing during specific nakaiy and planting crops during others, islanders have enough food to eat throughout the year; by thatching homes at the beginning of the wet season, palm fronds are given time and the necessary water to regrow. The extension of nakaiy to include a social dimension further builds resilience against the natural environment. By encouraging engagements during Adha, a rainy fortnight with poor fishing, ceremonies occur when fishermen have already pulled their dhonis out of the water. Rituals like the cutting of potato hairs, which are eaten in a stew during a wet season festival, ensure that all families in the island communities sow crops annually to participate in the rich celebrations. These celebrations and rituals help bind the nakaiy to community practices that enrich life and incentivize participation in natural resource cultivation and management.[10] Further, nakaiy are common, shared terminology in a country whose disparate geography offers little sense of Maldivian national identity. Seen through this lens, nakaiy act as a unifying framework in a geographically disparate nation with few commonalities except the language (Dhivehi), the climate, and the omnipresent ocean.

Holding a razor blade flush to an albacore's pectoral fin, Pradeep cuts into its flesh and methodically fillets the tuna. The central market in Malé is housed in an aluminum warehouse on a concrete pier. The market was built on land reclaimed from the sea, with sand dredged from a nearby lagoon. Like the rest of the capital, it is shielded from the ocean by a pile of concrete tetrapods that protect against storm surges.[11] The tetrapods were installed after a 1987 tidal surge that inundated the capital, the same year that Pradeep says the monsoon's arrival shifted for the first time in his lifetime.[12] Since then, the previously consistent weather patterns have become more unpredictable. "Hulhangu's storms used to stop fishing come April. Last year we stopped in July but the catch was bad. The storms came too late but they came quickly." Ali Rishman joins in: "I've seen it myself. It used to be that when the northeast monsoon began blowing, fish would jump off the reef and onto the islands. But that's all gone now."

The fishing industry has suffered as a result of climate change, with the formerly predictable nakaiy growing more erratic as the overfished stocks are gradually depleted. As the stability of the microseasons is weakened, nakaiy may become irrelevant and recede from the national lexicon to the domain of elderly fishermen, relegated to cultural lore of a bygone age. The erosion of legible and predictable patterns in the natural environment exposes the fragility of human society and its inability to control the climate. As their current definitions, intact for generations, become irrelevant and unintelligible, nakaiy risk fading as a Holocene luxury. Similarly, the relationship between the fishermen and the sea as well as the social interactions facilitated by nakaiy are unraveling. As their language becomes untethered from environmental reality, the twenty-six nakaiy must be transformed and redefined, or risk being archived as outdated vernacular.

The adoption of nakaiy as a loanword offers a proactive emphasis on local place identity that is critical to adapting to climate change. Nakaiy are not moment-to-moment Doppler radar measurements; nor were they created spontaneously. The form of twenty-six periods, their specific names, and the traditions surrounding nakaiy are all the result of intergenerational communal collaboration. In this way, the details of nakaiy are somewhat arbitrary: the tides, monsoons, and currents existed long before they were enlisted to describe discrete periods of weather. Nakaiy can be tailored to any community interested in the process of developing a calendar of hyperlocal microseasons.

Creating nakaiy outside of the Maldives will require similar processes of collaboration to generate local observations and knowledge of wind patterns, soil moisture, atmospheric pressure, aviary migrations, and other factors that are essential for comprehending the effects of climate change on communities of place.[13] If nakaiy are adopted and developed by communities at the town or city level, these place-based microseasons can simultaneously act as a baseline for variability in climate and drive adaptation to build resilience to climate change.[14]

Climate change presents a global collective action problem. Its effects seem intangible and irrelevant until they are considered at a scale at which the impacts can be felt, suggesting that localism is likely to be the required modus operandi of all communities in responding to climate change.[15] In a world with increasingly unpredictable weather, heightened self-sufficiency may be critical, necessitating a sharp decentralization from the world we now inhabit toward shorter distribution lines and more productive local economies.[16] This decentralization includes a shift toward community organizations and cooperatives that protect both producers and their surrounding ecosystems, encouraging local stewardship and investment, both in time and financial resources.[17] Beginning this transition to localism now ensures that grassroots movements will be keystones of communities and ease the transitional pains of shrinking economies to a low-energy and low-consumption future.[18]

The creation of local nakaiy would represent a conscious, collaborative shift toward localism. As such, the detailed and specific observations of weather patterns and abiotic factors would have to be spearheaded from within each community. Involvement in local social and political groups, such as those required to create nakaiy, are shown to be consistent and significant indicators of ties to local places.[19] Place identity is not static but rather a continuously constructed dialogue, in which "emotional attachment to places . . . develops through direct presence and activity at a locale."[20] The rigorous localism required to construct nakaiy offers a way for all citizens to engage with the world immediately around them. While fishermen, loggers, or others who are directly dependent on the extraction of natural resources may be most attuned to the climate, any group of interested individuals who share a landscape can measure temperature, precipitation, sunshine, wind chill, air and sea currents, soil acidity, visibility, glacial and permafrost melt, pollen release, or any of the hundreds of other factors that can contribute to a robust understanding of the local climate. High school science students, hunting clubs, environmental groups, nonprofit organizations, town governments, and senior citizens are all capable of engaging in the kind of grassroots citizen science initiatives that the development of nakaiy would require. Environmental artists, avid diarists, and outdoor bench enthusiasts can all add rich details to the community record. To validate citizen scientists, local universities and community colleges may be enlisted for access to scientific tools or previous experiments, while the Internet offers historical climate data as well as real-time updates from weather stations across the globe. Following the scientific method, after data are initially collected, trends and anomalies must be determined and explained. Group analysis, whether by extant organizations or a community task force dedicated to localized microseasons, would play a key role in determining the distinct and legible nuances of local climates. Questions about frost, sunlight, rainfall, and a host of other factors from the data collected can be used to devise a system of nakaiy that divides the calendar year into periods that are climatically notable. These periods may be given specific names or numerated by specific days of the year as start dates, but they must be broadcast, shared, discussed, and edited by an audience wider than their original creators.

In the Maldives, nakaiy are performative. They actively shape both the natural resource economy and the social lives of humans. In Western societies, even those whose landscapes are asphalt and concrete, communities can use nakaiy in similar ways. The timing of the first frost may end the community gardening season but signal the beginning of community-share agriculture from greenhouses. Weeks of highest rainfall can be proactively scheduled for indoor maintenance projects like cleanups and painting. Shade loving varieties can be planted with success in herb gardens and home vegetable plots in cloudy parts of town, perhaps in agreement with neighbors who live in sunnier areas who can trade apples for watermelons or carrots for corn. Local festivals celebrating the spring equinox, autumn harvest, clearest moon, or king tide are rooted in the observation of natural phenomena and are accessible to all. These events could be held to coincide with the beginnings of individual nakaiy, further embedding the hyperlocal into the community ethos and contributing to the rebirth of localisms and rituals anchored to the specific landscapes they celebrate.[21] These rituals would become embedded in cultural systems, encouraging greater investment in local communities and becoming "natural" within generations.[22]

Nakaiy are not stable ideas or definitions; they should not be thought of as judgments about the condition or desirability of a landscape but as collections of environmental knowledge that can help to bridge the divide between humanity and the natural world in a time of transformation. As such, communities must be able to evolve nakaiy by changing their definitions when observations of microclimate or weather pattern diverge from their functional description. These changes can be shared with scientists, politicians, and NGOs as documentation of environmental change over time, especially in societies that may require strategies for temporary resettlement or permanent migration in the future.

In the Maldives, former president Mohammed Nasheed has expressed fear that the nation's capacity to adapt might be overwhelmed and that attempts to preserve the islands from expected sea-level rise may be futile. He has publicly explored options for the transplantation of the entire population to neighboring nations. Nakaiy, which today are expressions of unity across disparate islands, may become nostalgic mementos for Maldivians displaced from their countrymen. The process of translating nakaiy to other locations could strengthen connections between neighbors, acquaintances, and arrivals displaced from their homelands. By bringing the soil, water, and sky surrounding local communities into focus, nakaiy might serve to enhance a sense of place and encourage investment in shared landscapes and social lives—gestures of unity amidst the upheavals of the Anthropocene.

Plant Time
ANOTHER PATH

NOTES

1. F. Abdulla and M. O'Shea, "English–Dhivehi Dictionary: A Guide to Language of the Maldives," Maldives Royal Family, 2005, http://www.maldivesroyalfamily.com /pdf/maldives_dictionary_1.0.pdf.

2. Jeroen Pijpe, Alex de Voogt, Mannis van Oven, et al., "Indian Ocean Crossroads: Human Genetic Origin and Population Structure in the Maldives," *American Journal of Physical Anthropology* 151, no. 1 (2013): 59.

3. Benjamin K. Sovacool, "Expert Views of Climate Change Adaptation in the Maldives," *Climatic Change* 114, no. 2 (2012): 299.

4. United Nations Development Program Maldives and Disaster Risk Management Team Maldives, *Detailed Island Risk Assessment in the Maldives: Executive Summary* (Malé: UNDP Press, 2007), ii.

5. Global Environmental Facility, *Project Identification Form: Integration of Climate Change Risks into the Maldives Safe Island Development Program* (Washington, D.C.: GEF Agency, 2009), 3.

6. Hassan Moosa and Geeta Anand, "Inhabitants of Maldives Fear a Flood of Saudi Money," *New York Times*, March 26, 2017, https://www.nytimes.com/.

7. Fishery and Aquaculture Organization, *Country Profile Fact Sheet: Maldives* (Rome: Fishery and Aquaculture Organization, 2009), 9.

8. Yameen Rasheed, "Dhoonidhoo Diaries: Part 3—Heat and Captivity," Daily Panic (blog), August 22, 2015, http://thedailypanic.com/.

9. Yi-fu Tuan, *Landscapes of Fear* (New York: Pantheon Books, 1979), 36.

10. Dolores LaChapelle, "Ritual Is Essential: Seeing Ritual and Ceremony as Sophisticated Social and Spiritual Technology," in *The Palgrave Environmental Reader,* ed. Daniel G. Payne and Richard S. Newman (New York: Palgrave Macmillan, 2009), 235.

11. Paul S. Kench, "Compromising Reef Island Shoreline Dynamics: Legacies of the Engineering Paradigm in the Maldives," in *Pitfalls of Shoreline Stabilization,* ed. Andrew Cooper and Orrin H. Pilkey (Amsterdam: Springer, 2012), 169.

12. United Nations Development Program Maldives and Disaster Risk Management Team Maldives, *Risk Assessment,* xi.

13. Dyanna Riedlinger and Fikret Berkes, "Contributions of Traditional Knowledge to Understanding Climate Change in the Canadian Arctic," *Polar Record* 37, no. 3 (2001): 319.

14. Fikret Berkes and Dyanna Jolly, "Adapting to Climate Change: Social-Ecological Resilience in a Canadian Western Arctic Community," *Conservation Ecology* 5, no. 2 (2001): 19; C. S. Holling, "Regional Responses to Global Change," *Conservation Ecology* 1, no. 2 (1997): 3; Gleb Raygorodetsky, "Why Traditional Knowledge Holds the Key to Climate Change," United Nations University, December 13, 2011, https://unu.edu/.

15. P. Newman, Timothy Beatley, and Heather Boyer, *Resilient Cities: Responding to Peak Oil and Climate Change* (Washington, D.C.: Island Press, 2009), 136; Riedlinger and Berkes, "Contributions," 28.

16. David Haenke, "Bioregionalism and Community: A Call to Action," Fellowship for International Community, 2005, https://www.ic.org/; Fred Magdoff, "An Ecologically Sound and Socially Just Economy," *Monthly Review* 66, no. 4 (2014): 1.

17. Naomi Klein, *This Changes Everything: Capitalism vs. the Climate* (New York: Simon & Schuster, 2014), 239; Pope Francis, "Encyclical Letter, Laudato Si', of the Holy Father Francis, on Care for Our Common Home," 2015 encyclical, http://w2.vatican.va/.

18. Tom Prugh, "Childhood's End," in *State of the World, 2015: Confronting Hidden Threats to Sustainability* (Washington, D.C.: Worldwatch Institute/Island Press, 2015), 129.

19. Lee Cuba and David M. Hummon, "A Place to Call Home: Identification with Dwelling, Community and Region," *Sociological Quarterly* 34, no. 1 (1993): 115.

20. Brian W. Eisenhower, Richard S. Krannich, and Dale J. Blahna, "Attachments to Special Places on Public Lands: An Analysis of Activities, Reason for Attachments, and Community Connections," *Society and Natural Resources* 13, no. 5 (2000): 423.

21. Robert D. Kaplan, *The Revenge of Geography: What the Map Tells Us about Coming Conflicts and the Battle against Fate* (New York: Random House, 2012), 35.

22. Erik Assadourian, "The Rise and Fall of Consumer Cultures," in *State of the World, 2010: Transforming Cultures from Consumerism to Sustainability* (New York: Worldwatch Institute/Norton, 2010), 3.

Pronunciation: pa t:wan

           (p̪ā fheṳ̈xn or Pa: tʰɯan in IPA)

Part of Speech: Noun ("wild forest")

Provenance: Thai

Example: "Why did the helicopter crash? It's the pa theuan!
It is full of ghosts—liver-eating ghosts. The kind that hop on
one leg. Old tree spirits like this one. The army thought they
could fly right over it. That's why it crashed."

Chao Mae Takhian lived in a tree trunk that shared her name. In fact,
she was, at times, the tree, and at times she was a beautiful woman with
flowing black hair and a penchant for sky-blue silk dresses. She also
had a penchant for young men, and, if she was too fond of or too angry
at one, she was known to send dangerous, even deadly accidents their
way so that she might "collect" them to live with her in her tree. These
included motorcycle accidents, surprise electrocutions—always a risk
given Bangkok's crumbling infrastructure and often-exposed power
lines—or sudden poisonings. At the same time, she could send gifts to
those who approached her in the right way—lottery ticket numbers, a
lover's affection, healing, these kinds of things.

    She was discovered lying in a bog in Ang Thong province, to
the north of Bangkok. She had appeared to a farmer in a dream and
indicated to him where she was hidden. The farmer was terrified but
thought it best to heed the dream. Just as Takhian had predicted, when

the farmer dug in the indicated spot, he found two portions of a gigan-
tic *Hopea odorata* tree buried in the black soil. One would imagine that
he would have found something, even without divine guidance—the
thick soil that formerly was the bottom of the Chao Phraya floodplain
and wetland, now cut and drained by canals and turned into rice paddy,
contains layer upon layer of remains.

Hopea odorata are tall and straight, with pale, smooth bark. They
stand out, even in the dense forests in which they normally grow.
Indeed, the comparison with an elegant (and even a little vain) noble-
woman is unsurprising. For those living on the forests' edge, the trees
are stark reminders of the size and scale of the forest. If Takhian's fear-
some image is any guide, they are also reminders that the forest contains both violence and beauty; in many parts of Southeast Asia, both
are aspects of the divine.[1] But here, rather than standing at the forest's
edge, Takhian had been cut down and tossed into a mire.

Bogs have always preserved biological material inside them—
trees, bodies, boats. In England, the discovery and removal of ancient
"bog oaks" were a particular sign of the difficult (but inevitable) progress when cultivating fenland (nutrient-poor alkaline marshes). In a
Lockean framework, the labor involved in pulling out these blackened
and soggy trunks added value to the land, becoming at times a part of
a Protestant ethic and patriotic duty.[2] To pull out the dense and sodden
wood was one (laborious) step toward making fenland arable, toward
making it productive.

Thailand's own twentieth-century cult of development worked in a
similar manner to this notion of labor as a moral value. Technology—
especially foreign technology—was the key element that could turn a
floodplain that was essentially a lake during the wet season into productive land. Thai state propaganda promised that the country, given the
right technological inputs, was destined to transform a flooded, canal-
based economy and ecology into a rationalized one.[3] Two rice harvests a
year could be three. Houses built on piles (to avoid the yearly flood) and

with gaps in the walls and floor (for cooling breezes) could be sealed, concrete, air-conditioned ones. Even the Thai monarchy came under this cult of development. Bhumibol Adulyadej, king for most of the twentieth century, was most often depicted doing development work: in a blazer, sweat dripping from his nose, clipboard in hand, camera around his neck. While later in his career he preached his own doctrine of self-sufficiency, this too was one built on the transformation of nature. His palaces designed new agricultural inputs, new rationalized city plans, and new economic theories (even if they were often ineffectual). The transformation of the flooded landscape became, like the English fens, a fusion of religious and economic notions of progress, except the narrative was a Buddhist march toward enlightenment, one led by those possessing wisdom (*panya*) rather than a Calvinist notion of labor.[4]

Such a notion was always hierarchical.[5] Royal proclamations to live within one's means—the "sufficiency economy" (*sethakit pho phiang*)—applied primarily to the poor, and further to an idealized poor that bore little similarity to actual rural livelihoods.[6] The rich, and especially the well-connected and prestigious rich (the "good people," *khon dii*), had no such need to mind their consumption patterns. But there were older ways of living with nonhumans, before such managerial modes of being.

It is these prior ways of being that Takhian and those like her demonstrate. The waterlogged tree trunks were a sign of what had come before: large hardwood forests cut and sunk into black water, then unearthed during the course of transforming the land. In Bangkok, though, there were other such members of the pantheon of discarded wild things. A dead cobra (*chao mae joong ang*), for instance, haunted a highway where a field of tall grass originally stood, causing traffic accidents to unsuspecting motorists but accepting supplication from those in Bangkok's marginal economy.[7] In Chiang Mai, Thailand's second city, Grandmother and Grandfather Sae—cannibals who posed a threat to uninvited wanderers in the forest—send cool breezes down into the city after they (or, rather, mediums possessed by them) are given gifts in a yearly feast of raw buffalo meat and blood.[8]

These beings were from the wilderness, in the sense of pa theuan.[9] The Thai word *theuan* refers to wildness. A *pa*, a forest, can be a manicured, maintained wood behind a temple. But a pa theuan, a wild forest, is beyond the control of humanity. This wildness can refer to anything that escapes social control—unregistered firearms, the underground economy, a savage street fight. It refers to something that is not standing reserve (following Heidegger) to be marshaled and maximized for gain, but rather to a thing from another realm that must be handled delicately.[10] Takhian, the cobra, and Grandmother and Grandfather Sae are all beings from the pa theuan that must be contended with through propitiation, negotiation, engagement, and respect, thus transforming a dangerous thing into a partner. But in doing so, their fundamental otherness is maintained. One is never certain that a deal with Takhian will turn out in the correct way, just as one always engages with uncertainty when making a deal with the cobra.

It is this notion that I seek to contrast with the concept of "wilderness" in English. In a settler-colonial milieu, wilderness exists to be overcome, a challenge to be tamed. Even a national park has been mobilized for a particular goal: tourism, conservation, recreation. One conquers the wild in order to bring it into a rationalized order, to adapt it to human understanding and human control. But something that is theuan cannot be conquered, merely engaged with on its own terms. Humans must, in dealing with Takhian and others, momentarily give way to theuan logics, not vice versa. In this way, dealing with things that are theuan does not involve moving the nonhuman into the human realm, but rather inhabiting a shared realm, one that is always imperfectly understood and recognizes radically different others as co-creators.

The pa theuan—the inassimilable wilderness—is therefore agentive. It is a subject position with rights, desires, and politics. But here I depart from much of the current vitalist literature that seeks to understand nonhumans as actors—where soy, for instance, can be an agent able to be tried in court for murder.[11] This is not what I mean. There is a difference in saying that *Hopea odorata*—the trees—are actors and saying that Takhian—the tree woman—is an actor. To talk about trees as enabling certain kinds of sociality, or networks of roots and rhizomes enacting particular ecologies, would be to reduce Takhian to a root network, to transpiration and carbon, or to the political economy surrounding *Hopea odorata* extraction. This would be to extend the logic of a human world with its rules and ontology onto something radically different, something that is in its essence inaccessible. In short, the theuan, being not only nonhuman but inhuman, is unable to be inhabited and thus can exist on its own terms.

Human knowledge has limits, and beyond these limits exist other ways of being that are uninhabitable but nonetheless should be allowed to exist. We can only imagine imperfectly how Takhian sees the world, but we deny her the space to live at our own peril. In allowing her to exist in her own right, we keep the space open for new worlds.

Yet she has not been allowed to exist. It should be remembered that Takhian, the cobra, and other beings from the pa theuan emerge into our world transformed. They have died; they have been chopped down, run over, and destroyed by Bangkok's expanding reach. It is only through their exhumation that they reemerge like uncanny reminders that every city rests (uneasily) on the bones of what has come before. While their other worlds still exist, they do so in the shadow of developmentalist progress—in urban shrines, under motorways, on the sides of roads. The appearance of the pa theuan in the city, then, is a reminder that other ways of existing with nonhuman others once were possible but are now spectral. Their haunting suggests the possibility of another way of being and the limits of our understanding.

What can we learn from Takhian and her kind? There is a particular kind of arrogance in a developmentalist scheme of knowledge, one that posits that we have a master plan that we can and will enact in order to make the world a better place. The river can be dammed and produce electricity. The fenland can be made arable. New crops can produce more calories, and we can fit more people into less space. We can get more from the world, this way of thought tells us, and we can profit, make new toys, and fatten ourselves.

It is at this moment that, Takhian's devotees would say, we run the risk of falling victim to those that we have overlooked. Remember that Takhian, like the dead cobra, attacks victims via traffic accidents, poisonings, electrocutions, and other failures of modern infrastructure. But in acknowledging the pa theuan, in recognizing that we share a world that we incompletely know with other beings that we also incompletely know, is a call to humility. It is a call to slow down our logic and rationality; above all it is a call to stop seeing the world as a place full of potential profit.[12] Living with theuan beings is a reminder that we exist in relationships with things that lie outside of our frames of reference. While we must deal with such nonhuman actors, what they teach us is that we are not the only agents of change, and our world is not the only one.

NOTES

1. See Tony Day, *Fluid Iron: State Formation in Southeast Asia* (Honolulu: University of Hawaii Press, 2002).
2. Richard D. G. Irvine, "East Anglian Fenland: Water, the Work of Imagination, and the Creation of Value," in *Waterworlds: Anthropology in Fluid Environments,* ed. Kirsten Hastrup and Frida Hastrup (New York: Berghahn Books, 2016), 23–45.
3. See James Scott, *Seeing Like a State: How Certain Schemes to Improve the Human Condition Have Failed* (New Haven, Conn.: Yale University Press, 1998).
4. See Daena Funahashi, "Rule by Good People: Health Governance and the Violence of Moral Authority in Thailand," *Cultural Anthropology* 31, no. 1 (2016): 107–30.
5. See Eli Elinoff, "Sufficient Citizens: Moderation and the Politics of Sustainable Development in Thailand," *POLAR: Political and Legal Anthropology Review* 37, no. 1 (2014): 89–108.

6. See Andrew Walker, "Royal Sufficiency and Elite Misrepresentation of Rural Liveli-hoods," in *Saying the Unsayable: Monarchy and Democracy in Thailand*, ed. Søren Ivarsson and Lotte Isager (Copenhagen: NIAS Press, 2010), 241–66.

7. See Andrew Alan Johnson, "Naming Chaos: Accident, Precariousness and the Spirits of Wildness in Urban Thai Spirit Cults," *American Ethnologist* 39, no. 4 (2012): 766–78.

8. See Andrew Alan Johnson, *Ghosts of the New City: Spirits, Urbanity and the Ruins of Progress in Chiang Mai* (Honolulu: University of Hawaii Press, 2014).

9. The Thai word is "เถื่อน" and is impossible to write using Roman characters, as it incorporates tone and a diphthong that does not exist in English. I Romanize this vowel as "eua" here, but it might be written "tuan" in the International Phoenetic Alphabet or "thüan," following a German pronunciation. Imagine pronouncing the English dipthong "oo-a" while smiling broadly. Other, related, meanings of the word *theuan* that I do not address here include "fake," "illegal," or "uncontrolled."

10. Martin Heidegger, *The Question Concerning Technology and Other Essays* (New York: Harper Torchbooks, 1977).

11. For vitalist literature, see Jane Bennett, *Vibrant Matter: A Political Ecology of Things* (Durham, N.C.: Duke University Press, 2010). For trying soy as an agent, see Kregg Hetherington, "Beans before the Law: Knowledge Practices, Responsibility and the Paraguayan Soy Boom," *Cultural Anthropology* 28, no. 1 (2013): 65–85.

12. Isabelle Stengers, "The Cosmopolitcal Proposal," in *Making Things Public: Atmo-spheres of Democracy*, ed. Bruno Latour and Peter Weibel (Cambridge, Mass.: MIT Press, 2005), 994.

Pachamama

Miriam Tola

Pronunciation: potch-a-mama (pɒtʃ:ä:mɒm:ə)

Part of Speech: Noun

Provenance: Aymara and Quechua

Example: "Water cannot be bought or sold, or subjected to market logic because water is a vital part of life: it is the blood of the pachamama."—Pablo Mamani Ramírez, 2004

Along the Andean cordillera, from Ecuador to northern Argentina, Pachamama is the name for capricious earthly forces embodied in rocks, rivers, and mountains. Relegated to the realms of religion and folklore by Western modernity, this powerful earth being remains an important presence in the everyday life of Andean indigenous communities.[1] In Ecuador and Bolivia, both plurinational states with highly politicized populations, Pachamama has become the subject of legal rights. Indigenous movements, in collaboration with environmentalists, have played a crucial role in jump-starting these reforms. A close look at states' and activists' visions of Pachamama, however, reveals remarkable divergences. Even the most progressive Andean governments have embraced an extractivist model of growth and claimed the sovereign right to decide how to use the so-called gifts of Pachamama, including water, gas, oil, and lithium. Social movements, in contrast, oppose state-led extractivism because it threatens the existence of Pachamama and indigenous ways of living. They contend that Pachamama, the vital energy that makes human life possible, should not be reduced to an economic asset.

194

What is at stake in these struggles is not just access to natural resources but the relationship between nature and society. As the tensions over Pachamama concern the possibility of creating alternatives to the capitalist commodification of nature, they matter far beyond the Andes. Usually translated as "Mother Earth" or "Earth Mother," Pachamama has long been associated to femininity. Because of this, issues of gender and sexuality play a significant role in the political conflicts unfolding in Latin America. As a loanword, Pachamama illuminates linkages between colonial history, extractivism, and gender. By troubling both the notion of the planet as repository of resources and the feminized trope of Mother Earth, it points to ways of inhabiting the earth otherwise.

Traces of Pachamama as a powerful other-than-human being embodied in the landscape emerge from pre-Hispanic archeological records. The Inca interacted with Pachamama through a range of agricultural and architectural techniques that integrated features of the landscape into the built environment.[2] The outcomes of these relational practices were uncertain and could bring prosperity as well as destruction. An unpredictable, even terrifying entity, Pachamama demanded attention, cajoling, ritual practices, and sacrifices. When treated with respect, the earth could respond with abundant harvests. Failure to pay proper attention to Pachamama, however, could lead to arid soils, illness, and even death. Although capable of generating life, the pre-Hispanic Pachamama could hardly be described as a benevolent, all-giving mother.

The writings of Spanish chroniclers provide vivid accounts of early European encounters with Pachamama. These texts describe the Andean earth being in terms that closely resemble Christian notions of proper femininity. Around 1575, Cristóbal de Molina, a Cuzco-based preacher and observer of Inca rituals, published a book detailing offerings and prayers to Pachamama.[3] The Inca, he claimed, celebrated Pachamama as their mother and as the mother of fire, corn, and seeds. The prayers translated by de Molina convey the image of Pachamama as a maternal figure capable of holding her children close to her, in calm and peace. The nurturing Pachamama also appears in the writings of Alonso Ramos Gavilán, a missionary dispatched to Peru in the early seventeenth century. In offering sacrifices to the earth, he argues, the natives ask Pachamama to respond as a good mother nourishing her children. Spanish chroniclers define Pachamama through the reassuring qualities that characterize the Virgin Mary. They assimilate the indigenous entity to the benevolent mother of the Christian tradition, the quintessential figure of sexual purity and domesticity. These European texts lay bare colonial attempts to incorporate indigenous belief systems into the Christian worldview.

The eighteenth-century Bolivian painting *Virgen del Cerro*, now in the collection of the Museum Casa Nacional de Moneda in Potosí, offers a striking visualization of such process of incorporation. In the anonymously created portrait, the Virgin's dress has the shape of a mountain: the notorious Cerro Rico of Potosí, which was home to the vast silver veins that fueled the Spanish colonial enterprise for over two centuries. The Holy Trinity holds a crown above Mary's head. The European powers, the Catholic Church, and the Spanish king appear at the feet of the mountain. They focus their attention on a blank globe, ready for colonization and conquest. The Virgin's richly decorated gown overlaps with the Andean landscape, as if Mary's body is mapped onto Pachamama and other Andean earth beings. Several other paintings of the same period show Mary as a mountain, thus indicating that this was an established iconography. Nevertheless, in spite of the identification between the Virgin and Pachamama, divergences remained significant for Andean communities. Although linked to generative powers, the pre-Hispanic Pachamama was not "virginal, chaste, or pure."[4] To European eyes, this entity and the indigenous women related to it suggested lust, lasciviousness, and moral chaos. By assimilating Pachamama to the Virgin, the Europeans attempted to bring under control a being and a worldview irreducible to Western hierarchical dichotomies of man/

woman, generativity/death, and purity/contamination. This history continues today. Echoes of colonial translation reverberate in the contemporary conflation between Pachamama and Mother Earth.

From green consumerism to climate justice movements, Mother Earth has been a ubiquitous presence in current evocations of the planet. In writings and public speeches, Pope Francis has invited Catholics to preserve Mother Earth in the face of planetary crisis.[5] German chancellor Angela Merkel has pledged to protect "our Mother Earth" from Donald Trump's disastrous environmental agenda.[6] Corporations and other neoliberal actors evoke the nurturing planet to make claims to sustainability even as they engage in environmentally destructive practices. Referring to a supposedly harmonious community of interconnected beings, this trope suggests the possibility of returning to an original unity with nature. Mother Earth is frequently presented as a fragile woman in need of saving, a vulnerable gendered subject that demands protection. Although evoked by anarchist Emma Goldman in the early twentieth century and more recently by environmental justice activist and ecofeminist Vandana Shiva, this metaphor has been widely contested by feminists of various stripes.[7] The trope of the earth as matriarch perpetuates the old habit of feminizing nature and confining women to the realms of reproduction and care. As ecocritic Stacy Alaimo points out, invocations of Mother Earth unwittingly render polluters as "unruly children," turning a systemic problem into a personal one.[8] Even more, the image of the benevolent earth fails to capture the nonlinear dynamics of socionatural systems in late capitalism. It obfuscates the turbulent complexity of geological, biological, and chemical processes that human action affects and amplifies but is not capable of controlling. Still, the gendered metaphor of Mother Earth has been embraced by environmentalists, both radical and mainstream, and it is widely associated with indigenous modes of living in relation to the land. In Latin America contexts, "she" is often conflated with Pachamama, with paradoxical effects.

In recent years, Pachamama has become increasingly visible in Latin American politics. Particularly in Bolivia and Ecuador, its new relevance is linked to the partial state recognition of indigenous worldviews. Ecuador and Bolivia have adopted the indigenous notion of *buen vivir* (from the Aymara *suma qamaña* and the Quechua *sumaq kawsay*) as the organizing principle of new constitutions approved in 2008 and 2009, respectively. Translated in English as "living well," *buen vivir* indexes indigenous alternatives to the Western model of development that prioritizes economic growth. In contrast to Western celebrations of individual agency and private property, *buen vivir* emphasizes the dependence of human communities on the ecologies of specific places on earth. According to Ecuadorian indigenous activist Monica Chuji, *buen vivir* demands that nature is seen not as a "factor of production" but as "an inherent part of the social being."[9] Informed by the demands of indigenous and popular movements, the 2008 Ecuadorian constitution builds on *buen vivir* to declare that Pachamama "has the right to integral respect for its existence and for the maintenance and regeneration of its life cycles." This relationship with Pachamama undergirds a "new form of citizenship coexistence, in diversity and harmony with nature." Similarly, in Bolivia, the government of Evo Morales, the first self-identified Aymara president and a former leader of the coca growers' union, supported the introduction of the 2012 Framework Law of Mother Earth that confers legal rights to ecosystems. Morales bolstered his reputation as champion of the rights of Pachamama in a series of international gatherings. His cry of "Pachamama or death" resonated deeply among the participants of the People's World Conference on Climate Change, which was held in the Bolivian city of Cochabamba in 2010. The same year, speaking at the general assembly of the United Nations, he boldly claimed that the flourishing of Mother Earth stands in direct opposition to capitalism.

Both in Ecuador and Bolivia, however, the conferral of rights to Pachamama has been fraught with contradictions. Despite political rhetoric to the contrary, the administrations of both plurinational states have prioritized economic development over *buen vivir*. Operating within the constraints of global capitalism, they rely on the neoextractivist model widely pursued across Latin America as a source of revenue for redistributive policies. Extractive industries, often funded through state–private partnerships, have been expanding; for example, the land areas conceded to gas and oil companies have increased exponentially.[10] In 2011 the Bolivian police violently raided an encampment of indigenous groups protesting the construction of a 182-mile highway cutting across a national park and indigenous territory known as Tipnis.[11] The project, canceled by the Bolivian government after the protests, was revived with a law passed in 2017. In 2013, Ecuadorian president Rafael Correa scrapped the Yasuni initiative that promised to keep considerable reserves of oil in the ground of the Yasuni Park, one of the most biodiverse areas of Amazonian Ecuador. In the original plan, Ecuador would abandon extractive plans for the area, reducing carbon emissions in return for donations from carbon-rich countries. Since only a small fraction of the promised compensations were donated to Ecuador, the government opened the area to oil exploration. A wave of protests followed Correa's announcement. The activists invoked the rights of Pachamama and cited the Ecuadorian government's decision as unconstitutional. Ongoing protests in Ecuador and Bolivia show a widening rift between extractivist agendas and social movements pushing against the use of land as a mere economic asset. After much initial excitement about the potential of progressive governments, indigenous organizations and intellectuals have been criticizing their hypocritical use of *buen vivir* and the rights of Pachamama.

In Bolivia, the mobilization of Pachamama in the project of state consolidation has taken a strikingly gendered dimension. In the international scene, Evo Morales presented himself as the guardian of Mother Earth, a feminized planet in need of protection against Western capitalism. At home he celebrates Pachamama for providing the country with cheap gas and other resources. Engaging in colonial translation, the Bolivian government thus participates in the rendering of Pachamama as the all-giving Mother Earth. The state frames the Andean earth being as a gendered subject of rights while at the same time it asserts a sovereign role in managing "her" mineral gifts. Bolivian feminists have exposed the contradictions of the state's mobilization of Pachamama. Aymara mestiza sociologist Silvia Rivera Cusicanqui argues that rather than enabling a radical shift in the relation between humans and nature, the government uses a Pachamama-ist rhetoric to sugarcoat neocolonial forms of appropriation of resources.[12] The feminist decolonial organization Feminismo Comunitario explicitly rejects the association between Pachamama and Mother Earth.[13] According to the group, the insistence on Pachamama as provider of resources available for extraction is problematic in many ways. The state, they contend, has turned the Andean being into an all-giving female body, and by association it has naturalized the reduction of Bolivian women to their reproductive capacities. This is especially troubling given Bolivia's restrictive abortion regime. Further, Pachamama should not be conflated with Western understandings of nature as economic asset. As the communitarian feminists put it, "While people are part of the Pachamama, the Pachamama does not belong to anyone."[14] Such perspectives are in line with the indigenous struggles in defense of Tipnis and Yasuni Park. Although the concerns of feminist and indigenous activists are not always the same, in Bolivia, they often coalesce in reclaiming Pachamama as a figure incompatible with the government's developmentalist agenda.

Pachamama is both ancestral and contemporary. Its temporality defies the Western linear conception of time and the myth of the vanished Indian. Pachamama is ancestral in that it belongs to past indigenous worlds devastated by European weapons, diseases, and trade. It is contemporary because, together with indigenous people, it has survived conquest and participates in struggles against the ongoing neocolonial dispossession that fuels global capitalism. We should not forget that the modern world was built through the plundering of Amerindian land-scapes, and that for the indigenous populations of South America, the arrival of the Europeans meant confronting extinction. Those who survived genocide had to invent new modes of living within colonial states that treated them as cheap labor, insignificant minorities, and exotic ornaments packaged for tourist consumption. In this sense, indigenous people are "veritable end-of-the-world experts."[15] Pachamama's persistence in reinvented indigenous worlds draws attention to what it is like to live in damaged landscapes.[16]

As a loanword, Pachamama offers a counterpoint to the Anthropocene, which many have embraced as the name for an epoch of human-made planetary change. The term *Anthropocene* presents the human species as a unified whole, thus effacing the difference between those responsible for the environmental crisis and those at risk of suffering its harsher consequences. Pachamama, in contrast, throws into relief the histories of colonial violence and capitalist dispossession that are deeply implicated in the uneven unfolding of ecological devastation. At the center of these histories there were Europeans, particularly white men, who considered themselves the measure of humanity and devalued indigenous people as less than human. The Anthropocene narrative focuses on the human as the primary agent of planetary change, simultaneously causing and remediating environmental collapse. British journalist Mark Lynas, for example, argues that "we" humans can still use science and technology to identify and solve environmental problems.[17] By drawing attention to unpredictable, even unruly, earth powers, Pachamama challenges the enthusiasm for technological fixes and the fantasies of controlling nature that underpin them. This loan-word calls for ways of living in which the notion of "progress," including technological progress, is no longer central.

Emerging out of Andean histories and struggles, the term Pachamama gestures toward the persistence of indigenous modes of living within and against the Anthropocene. Together with the Latin American feminists and indigenous activists who evoke it, it opens ways to conceive the earth, and its politics, otherwise. Pachamama is not as reassuring as the nurturing Mother Earth. It does not guarantee the return to a balanced planet. Rather, it provides glimpses of radically different modes of engaging gender, nature, and environmental politics.

NOTES

1. Marisol de la Cadena, *Earth-Beings: Ecologies of Practice across the Andean Worlds* (Durham, N.C.: Duke University Press, 2015).

2. Carolyn Dean, *A Culture of Stone: Inka Perspectives on Rock* (Durham, N.C.: Duke University Press, 2010).

3. Cristóbal de Molina, *Account of the Fables and Rites of the Incas* (Austin: University of Texas Press, 2011).

4. Veronica Salles-Reese, *From Viracocha to the Virgin of Copacabana: Representation of the Sacred at Lake Titicaca* (Austin: University of Texas Press, 1997), 38.

5. Pope Francis, *Laudato Si': On Care for Our Common Home* (encyclical), 2015, http://w2.vatican.va/content/francesco/en/encyclicals/documents/papa-francesco _20150524_enciclica-laudato-si.html.

6. Alison Smale, "Angela Merkel and Emmanuele Macron Unite behind Paris Accord," *New York Times*, June 2, 2017, https://nytimes.com/.

7. Stacy Alaimo, *Undomesticated Ground: Recasting Nature as Feminist Space* (Ithaca, N.Y.: Cornell University Press, 2000); Vandana Shiva, *Earth Democracy: Justice, Sustainability, and Peace* (London: Zed Books, 2005).

8. Alaimo, *Undomesticated Ground*, 174.

9. Monica Chuji, "Sumak Kawsay versus desarrollo," in *Antología del Pensamiento Indigenista Ecuatoriano sobre Sumak Kawsay*, ed. Antonio Luis Hidalgo-Capitán, Alejandro Guillén García, and Nancy Rosario Deleg Guazha (Cuenca, Ecuador: FIUCUHU, 2014), 231–33.

10. Emily Achtenberg, "Morales Greenlights TIPNIS Road, Oil and Gas Extraction in Bolivia's National Park," NACLA (North American Congress on Latin America), June 15, 2015, https://nacla.org/.

11. Silvia Rivera Cusicanqui, "Strategic Ethnicity, Nation, and (Neo)colonialism in Latin America," *Alternautas* 2, no. 2 (2015): 10–20.

12. Cusicanqui, "Strategic Ethnicity."

13. Feminismo Comunitario, "Pronunciamiento del feminismo comunitario latino-americano en la conferencia de los pueblos sobre cambio climático," May 2010, http://mujerescreandocomunidad.blogspot.com.

14. Feminismo Comunitario, "Pronunciamiento."
15. Deborah Danowski and Eduardo Viveiros de Castro, *The Ends of the World* (Cambridge: Polity Press, 2017), 108.
16. Anna Tsing Lowenhaupt, *The Mushroom at the End of the World* (Princeton, N.J.: Princeton University Press, 2015).
17. Mark Lynas, "A Good Anthropocene?," speech by Mark Lynas delivered at Breakthrough Dialogue 2015, Cavallo Point, San Francisco, Calif., Mark Lynas (blog), June 22, 2015, http://www.marklynas.org/.

Plant Time

Charis Boke

Pronunciation: plant time (plænt taɪm)

Part of Speech: Noun and adjective

Provenance: Some Western herbalists in North America

Example: In today's class, we'll head out to the garden. Everyone will choose a plant to sit with for a while, to get down to plant time before we start our work.

STORIES OF THE PRESENT: HERBALISTS PRACTICING

How can plants change minds? Herbalist students cultivate connections to plant time as part of their training in learning about medicinal plants and health, both human and ecological. They learn to embody human obligations to environmental others by "getting down to plant time." Sitting with plants in the garden shifts students' encounters with the world, as medicinal plants become sensible creatures and active collaborators. In this way, "getting down to plant time" is an embodied sensory practice. It takes time to learn and requires repetition. The iterative nature of many sessions of sitting with plants builds bodily attunements that enable attention and care across biologies.

I learned about the concept of plant time from a community of herbalists learning, practicing, and cultivating medicines at a school teaching Western herbalism, a broad subdiscipline of herbalism, in the rural Northeast of the United States. Teachers at this school use plant time as a practice to shift their own bodies' sensory attunements toward plants. Herbalists understand this vegetal mode of bodily presence as a first step in healing relationships between humans and plants. Building intimate, sensory relationships with plants enables herbalists to understand their own animacy—their own lives—as inextricably bound up with the plant-others of the world. This sensory partnership with plants teaches herbalists how to care for a world with beings, and needs, beyond the human.

As a loanword, plant time aims to make space for cross-species connections, communications, and modes of animacy. It is a bodily practice that enacts a shift in the sensory attunements of the herbalists' body as she approaches vegetally paced life. Plant time honors the difference between human time and vegetal temporalities—the cycles of germination, growth, maintenance, seeding, and dormancy or death that are months, years, or decades long. If Western herbalists articulate their own experience of a world of boundaries, borders, and notions of linear progress as a challenging one, characterized by the unstoppable cascade of calendar time, plant time describes the slower, cyclical time-lives that our vegetal kin inhabit. Anthropologist Natasha Myers calls this process of paying attention to the material entailments of other elements of the world "attunement."[1] This era we live in, called by some late capitalism and by others the Anthropocene, comes with more than its fair share of environmental, economic, interpersonal, and bodily challenges. The confluence of these challenges demands that we learn to *pay attention* in new terms. Our attentions are always shaped by the materials of the world, whether proteins, plants, or protozoa.[2] By attempting to move into plant time, herbalists construct a fleshy, embodied mode of attuning to such social and material entanglements. Herbalist attunement, the process of sitting and learning to attend to plants and their seemingly small worlds, and to what we experience as that vegetal pace distinct from human time, is the first tier of learning how to relate across species difference.

As an herbalist practice, "getting down to plant time" involves sitting with a growing, living medicinal herb—for ten minutes, three-quarters of an hour, until dusk. This practice shapes how herbalists learn to pay attention to the lives, cycles, connections, ecologies, and capacities of plants. Practitioners are motivated first by a sense that medicinal plants have much to offer, and next by a sense that medicinal plants have something to teach humans about plant life, and life in general. Not only does vegetal biochemistry offer medicine for humans (or poison, depending on the dose) but plants also teach herbalists a different way to experience being entangled in place and to be responsible to other critters. Entanglement and responsibility can be both concepts and practices, and they manifest in both ways for herbalists' practices with plant time. Physicist Karen Barad helps me to think about what herbalists are doing with plant time by using these terms in both ways: as concepts for understanding human presence in the world, and as practices for living more fully into our presence. After Barad, I might call plant time a process not just of becoming intimate with plants and building sensate bodily relationships sensitive to communication across species, but also of becoming animate with plants.[3] Agency is not a property of a bounded individual but an active process, produced in collaboration between multiple sorts of beings. Coming into being, and into capacity to act in the world, is always something that must, *must at a quantum physical level*, she says, happen *together* with the other stuff in the world. Action is always intra-action, and agency is always intra-agency. Action and agency cannot happen without mutual relationships among different kinds of materials and actors.

These entanglements mean that our actions and agencies are always in implicit relations of responsiveness to each other, entanglements Barad approaches with her theory of agential realism. Barad suggests, along with science studies scholar Donna Haraway, that we must learn to attend to our *response-abilities*.[4] Working with this idea of response-ability, I suggest that part of what herbalists do when they get down to plant time is learn how to become coanimate with vegetal others, coordinating across complexly embodied and divergent experiences of temporality.

Plant time practices attune human bodies and senses to be present to what is and what could be. Herbalists try not to appropriate from Indigenous cultural practices that already understand plants as kin and as worthy of respect, attention, and relationship. Herbalists' efforts to be inspired by Indigenous ways of engaging with plants without engaging in yet more colonial violence trace a tension around what kinds of practices can be taken up, how, and by whom. The "slow violence" of colonial imperialism has made some kinds of knowledge available to white herbalists by putting the idea of "traditional ecological knowledge" on a pedestal.[5] The confluence of these tensions, histories, and sensations lays the groundwork for herbalists' movement toward plant time–based communications with plants. Herbalist practices with plant time try to "unsettle," in Eve Tuck and K. Wayne Yang's felicitous term, a commonsense white relation with environments and plants.[6] They seek to deindustrialize and un-rush human experiences of the world, to carve out a space and time where human sensory experience can shift into another register, dropping its addiction to the calendar and the clock. This is what plant time might teach us: it works against a notion of an Anthropocene that centers specific kinds of human agency, power, and experience as isolated processes. Herbalists building relationships with plants through plant time work to re-embed human sensibilities within our ecological realities.

INSTRUCTIONS FOR PLANT TIME

Sally seats us on her sloping lawn in a warm late June patch of sunlight. She's here with twenty-five herbalist students to share practices and perspectives on getting to know plants as lively critters we live along-side, critters with whom we can relate. She says,

> We live in a society where it's easy to forget that we are
> connected with other parts of the world. With plants, with
> animals, with the Earth. This disconnection, our sense of
> disconnection, is at the heart of many ailments we see today.
> But it's not necessary—this disconnection is part of a system
> that is broken. I want to tell you, to remind you, that it's our
> birthright to be in connection—not separate from nature,
> but part of it.

Looking around the circle, she makes eye contact with each student.

> I have a friend, another herbalist, who has been teaching
> about plants and herbal medicine for many years. He had
> a practice in his workshops of asking people, as they open
> the workshop, to touch the earth. Over many years, every-
> one would close their eyes and touch the ground or the floor
> beside them. Only one person ever touched themselves.

She pauses. There is a silence around the circle. Some people nod in agreement; some eyes widen in surprise and realization.

> We are all a part of the earth already. This practice of sitting
> with plants has helped me recognize the value of vulnerability.
> It's easier to be vulnerable with a plant, with a tree. We
> still have to work to listen to what they share because we
> are so accustomed to being defensive. But as you recognize
> those defensive parts, it becomes easier to hold your own
> power in a soft and gentle way, to open, soften, make space.
> With this practice, we can respond differently to what arises
> in the world.

Sally gave the students an assignment that day: sit with plants at least four times before the next class. The instructions for how to encounter and interact with plants that students sat with were as follows:

With the plants:

- Introduce yourself.

- State your intention clearly—why are you here with this plant?

- Ask for what you need in another person, in a plant or tree, in the land.

- Practice being present with yourself.

- Pay attention to the shifts that happen in your body when sitting with a plant.

Herbalists like Sally frame plant time as a sensate bodily state that involves slowing heart rates, attention to breath and the immediately present green growing world, greeting and thanking a plant, and "listening" for what it might have to "say." I put these terms in quotation marks because they are imperfect representations of herbalists' intended meanings: "listening" and "saying" represent the ways that humans sense plant communications viscerally and intuitively, with eyes and with ears. Biologists have recently demonstrated that arboreal entities such as trees, herbaceous shrubs, and grasses communicate with one another. For the most part, these communications are accomplished through pheromones released into the air, or through connections between plants in the soil facilitated by signals conveyed through mycorrhizal networks of fungus (aka the "wood wide web").[7] Some communications happen only between individuals of a particular species clustered near one another. For instance, a deer gnawing at the bark of a poplar tree can cause the tree to release a pheromonal signal that travels across the grove of trees of the same species, triggering a surge in the production of certain browse-deterring chemicals in the bark of its neighbors. Other cross-species communication relies on mycelial fungal networks to facilitate the exchange of resources—water and nutrients—among the root clusters of plants connected by that network.[8]

Native, Indigenous, and First Nations peoples have long known that plants, as well as other critters, have the capacity to communicate. Anishinaabe scholar Wendy Makoons Geniusz has pointed out that "traditional knowledge" is imagined as both natural and idealized, while at the same time it is positioned as "primitive." She suggests this is something that can only happen as a result of the ways that settlement and colonization unfolded in what is now called North America.[9] Thus, while biologists' "discovery" of plants' abilities to exchange information and resources thrilled the world of Western biosciences, it was not news to everyone. The fact that plants communicate *differently* than humans should not be a barrier to understanding that they are capable of communication. As biologist Robin Wall Kimmerer puts it,

> [Science concluded] that plants cannot communicate because
> they lack the mechanisms that *animals* use to speak. . . .
> There is now compelling evidence that our [Indigenous]
> elders were right—the trees *are* talking to one another. . . .
> The mycorrhizae may form fungal bridges between
> individual trees, so that all the trees in a forest are connected.
> These fungal networks appear to redistribute the wealth of
> carbohydrates from tree to tree. . . . They weave a web of
> reciprocity, of giving and taking.[10]

Here Kimmerer describes a materially grounded approach to understanding plant communications. For her, "materials" include the interactions of molecular components such as carbohydrates as much as large-scale critters such as trees and deer. In order to think about what is involved in communication, she focuses on carbohydrates and phero-mones. Though those material components might be microscopic in size, they are more tangible to the average reader schooled in and com-mitted to the logics of Western biosciences than loose concepts such as "plant speech," "energy," or "intuition." Each individual plant *is* because many plants and other critters *are*—in what Barad might call interagency, and what I think of as becoming animate together. Herbalists under-stand plant time as a mode of bodily attention that can help humans develop similar relational capacities. It blends biological frames with human sensory experience to posit that plants choose when, how, and with whom they communicate.

BECOMING ANIMATE

What might it mean for humans to become animate with plants? Herb-alists contend that feeling our way into presence with plants can help us root ourselves, in the mess and muddle of our humanity, into the grow-ing world of the planet we live with, whether green and lush or sandy, stark, and succulent. The world requires of us, as it always has, that we pay attention—to our bodies, to the experiences of other humans, to the cycles of living and dying of the critters who share the world with us.

Human bodies, like plants, offer evidence for what is occurring. We hurt, we grieve, we startle, we anger, we fear, we sicken and die, we shout with joy and welcome. If plant time's bodily practices teach us how to pay attention, then we may be able to tune in more clearly to the ways that these reactions demonstrate our ability to become animate with plants and the rest of the world. Barad's notion that agency only emerges in relationship, in intra-action, grounds my artic-ulation of plant time as a bodily method for learning how we become animate, intentionally, with plants. Our bodies can teach us how to open possibilities beyond what we already know to be failed practices—extraction, colonization, a culture of control and power and profit. For the herbalists who use it, becoming attuned to plant time is a learned shifting of bodily sensibility. Changing corporeal orientations to others—starting at the intimate and moving outward into reshaping community, polity, and ecology—can change the world as we know it, and plant time changes how we know the world. This is what the plants can teach us.

STORIES OF THE (NEAR) FUTURE

The question plagued her as she wandered around the encampment, unable to work. She hadn't been sleeping well, waiting for what she thought was coming next, and what she knew had already happened. Years earlier her teacher had told her that without the ability to listen to plants, whatever that meant, humans' lives would end—or their meaning might. The teacher had demanded that she remember the right questions. What kind of kin are plants? She turned toward the edge of the cleared area, walking up to a small flowering tree. A few offerings of shell and glass and bone and metal, remnants from the lost twentieth-century world she barely recalled, rested on a sun-heated boulder beside the tree. She put her palm to its trunk, sharing her name silently, thanking the tree for its presence, asking to sit with it. Settling on the broken ground at its roots, she leaned on the trunk and closed her eyes. The sounds of the camp blended with the sounds of its beyond, past the tree. What kind of kin are plants? I can't remember. Tree, can you help? She paused, remembering to listen.

Friends. We can be friends. Still.

She opened her eyes. The sun shone through the blowing patches of silica particulate, glinting and shimmering. A bird—she thought maybe it was a robin, but she'd never been sure what they sounded like—croaked through its soot-scarred throat. As she looked toward the center tent where the circle would shortly start, she noticed more. Her belly felt settled, her brain-body clear. She had only asked and heard those few things in words, probably not the tree's words—her own translation. But still, she felt clearer than when she had sat down. She felt ready for the questions the council would ask each other— ready to ask them to remember their training, ready to ask them to bring that kind of attention and clarity back to their conversing. She smiled, palmed the tree in thanks as she stood, and pulled a few strands of her hair to leave with the shells and metal and glass.

Sehnsucht
ANOTHER PATH

NOTES

1. See Natasha Myers, "Ungrid-able Ecologies: Decolonizing the Ecological Senso-rium in a 10,000 Year-Old NaturalCultural Happening," *Catalyst* 3, no. 2 (2017): 1–24. See also Natasha Myers, *Rendering Life Molecular: Models, Modelers, and Excitable Matter* (Durham, N.C.: Duke University Press, 2015); and Myers, "Sensing Botanical Sensoria: A Kriya for Cultivating Your Inner Plant," Centre for Imaginative Ethnography, n.d., http://imaginativeethnography.org/.

2. See Alex Nading, *Mosquito Trails* (Berkeley: University of California Press, 2014); and Eben Kirksey, *Emergent Ecologies* (Durham N.C.: Duke University Press, 2015).

3. See Karen Barad, *Meeting the Universe Halfway: Quantum Physics and the Entangle-ment of Matter and Meaning* (Durham, N.C.: Duke University Press, 2007); and "Intra-actions: Interview of Karen Barad by Adam Kleinman," *Mousse* 34 (2012): 76–81.

4. Barad, *Meeting the Universe Halfway*; Karen Barad, "On Touching—The Inhuman That Therefore I Am," *differences* 23, no. 3 (2012); and Donna Haraway, *Staying with the Trouble: Making Kin in the Chthulucene* (Durham, N.C.: Duke University Press, 2016).

5. See Rob Nixon, *Slow Violence and the Environmentalism of the Poor* (Cambridge, Mass.: Harvard University Press, 2013).

6. Eve Tuck and K. Wayne Yang, "Decolonization Is Not a Metaphor," *Decolonization* 1, no. 1 (2012).

7. See Robin Wall Kimmerer, *Braiding Sweetgrass: Indigenous Wisdom, Scientific Knowledge and the Teachings of Plants* (Minneapolis, Minn.: Milkweed Editions, 2015); and Sabine C. Jung, Ainhoa Martínez-Medina, Juan A. López-Ráez, and

María J. Pozo, "Mycorrhiza-Induced Resistance and Priming of Plant Defenses," *Journal of Chemical Ecology* 38, no. 6 (2012): 651–64.

8. Suzanne W. Simard, David A. Perry, Melanie D. Jones, David D. Myrold, Daniel M. Durall, and Randy Molina, "Net Transfer of Carbon between Ectomycorrhizal Tree Species in the Field," *Nature* 388, no. 6642 (1997): 579–82.

9. See Wendy Makoons Geniusz, *Our Knowledge Is Not Primitive: Decolonizing Botanical Anishinaabe Teachings* (Syracuse, N.Y.: Syracuse University Press, 2009).

10. Kimmerer, *Braiding Sweetgrass*, 20.

Pronunciation: chee (tʃi)

Part of Speech: Noun (uncountable)

Provenance: Chinese, mainly Confucianism and Daoism

Example: According to the United Biota ECOnomic Taskforce (UBET), the ECOnomic analytics for the last month show signs of sustained harmony and balance, as the flow of qi at the anthropic level continues to align with the flow of qi at the ecological level.

The qi between heaven and earth flows in order.
A rattled order brings chaos to the people. When the
yang remains eclipsed and unexposed, while the yin
overwhelmed and inescapable, earthquakes ensue.

——BOOK OF ZHOU, HISTORY OF NATIONS[1]

In the second year of the king of You's rule in the Zhou Dynasty (780 BC), three earthquakes jolted the Chinese nation. They all took place along major rivers—regions home to vast populations in the largely agricultural economy of the Zhou era. Reports of casualties kept escalating. The king of You summoned his chief historian, Father Boyang, to the imperial court and inquired about these earthquakes. Father Boyang made the comments found in the epigraph and candidly predicted the demise of the dynasty within the next decade. Nine years later, the king of You was assassinated, ending the nearly three-hundred-year Zhou Dynasty.

Father Boyang's observation was remembered not only for the accuracy with which it prophesied the demise of the Zhou Dynasty but also for being the first articulation of qi as a philosophical concept in ancient Chinese thought. His description of qi as a self-balancing order of universal being that regulates everything between heaven and earth has inspired generations of Chinese thinkers to further elaborate on its meaning.[2] It is a concept that has endured the test of time.

Why does the notion of qi warrant consideration as a loanword in the Anthropocene? Our present predicament has been characterized by philosopher Stephen Gardiner as the "perfect moral storm,"[3] marked by a plethora of interrelated environmental challenges. This "perfect moral storm" blows through familiar spatial orders (of cities, regions, and nations) and temporal scales (of days, years, and decades). It poses a major cognitive challenge for people to make sense of the conditions of the biosphere. In part, the "perfect moral storm" stems from the lost connection between human societies and the natural world—between culture and nature. From the food we put into our bodies to the gasoline we pump into our cars, elements of nature are so far removed from their origins that we hardly recognize them as such.[4] Moreover, the complex web of ecological relations is partitioned into separate domains of knowledge—culinary science, petroleum engineering, ecotourism management, and the like. The contemporary human experience is also conveniently compartmentalized into binaries: local versus global, city versus wilderness, private versus public, and, above all, us versus them.

Qi offers a way to reconstruct our ecological imagination for the Anthropocene. Two qualities of qi are particularly worth noting. First, it is universal. Qi describes an indiscriminate force that sustains all fauna and flora on earth as well as human communities. It creates an all-encompassing field of dynamism in which different forms of life connect, interact, and exchange qi. Second, qi is self-balancing. The well-being of life-forms does not depend on the growth of qi but on its yin–yang balance. Yin and yang describe ideal types of seemingly opposite ends of a spectrum, such as light and dark, cold and heat. Although they appear contradictory to human senses, they are interdependent at the general level of qi. In fact, they constitute a check on each other; too much of either yin or yang disrupts qi and threatens harmony. As Father Boyang put it, "A rattled order brings chaos." Chaos, which seems destructive at the moment of its occurrence, restores the yin–yang balance of qi in the larger scheme of things.

At the core of qi is an emphasis on connectivity. It embeds individual humans in the network of social relations as well as in the web of ecology. This conception echoes sociologists Michael Bell and Loka Ashwood's understanding of the environment as "the biggest community of all."[5] In this big community of farmers, blacksmiths, traders, beavers, elephants, pine trees, rainforests, volcanoes, and all other earthly beings, qi is the common source of vitality. Members of this community are brought together by qi.

Curiously, the flows of qi find their most elemental expression in the human body and soul. According to strands of ancient Chinese thought, qi is fundamental to the well-being of individuals, both physically and spiritually. The yin–yang balance in qi, which was the basis of Father Boyang's diagnosis of Zhou Dynasty's predicament, figures prominently in traditional Chinese medicine (TCM).[6] Qi is thought to flow in all bodily organs. The human body gains supply of qi not only through breathing fresh air but also by consuming different types of food. The *Book of Five Tastes of the Divine Pivot*, a Chinese medical classic compiled in the first century BC, notes that "half a day without grain diminishes *qi*, and a full day exhausts it."[7] However, the relationship between the amount of food consumed and the amount of qi gained is not linear. Some types of food contribute to qi of yin quality, which is calming and soothing to the human body. Examples include cucumber and pear.

Conversely, other foods, such as pepper and ginger, produce qi of yang quality, which is energizing and stimulating to the body.[8]

The principal TCM diagnosis of a bodily dysfunction, therefore, rests on an analysis of qi that circulates in the troubled organ or organs. Sickness is said to result from insufficient qi, imbalanced yin–yang, or a combination of both.[9] A common TCM treatment is to use particular types of herbs to help restore the abundance and balance of qi. However, that alone is hardly sufficient. TCM practitioners often give advice about diet, exercise, sexual activity, and a range of other aspects of life.[10] When TCM practitioners advise patients to refrain from disturbing qi, they refer not just to breathing air and digesting food (both are material manifestations of qi) but also the preservation of temperament, integrity, and honor (all immaterial forms of qi). An edgy person is described as having agitated qi, a lucky person managed qi, and an upright person squared qi. In this sense, physical well-being is simply an observable display of a healthy mind and heart, and vice versa. Seemingly disparate aspects of an individual's life are thus connected and thoroughly integrated thanks to qi. After all, the death of a person is announced as the moment when one swallows their last breath of qi.

Yet qi is not just a matter of private concern; it also figures prominently in social life. When people get along well, they are said to have a mass of congenial qi. When they don't, they are said to produce qi against each other. The same qi, in other words, works both ways: it can bring people together or pit them against each other. It is fraught with contradictions. Interestingly, a mass of congenial qi is shared, benefiting everyone in the group. When one produces qi against another, however, the most immediate consequence is that the producer of qi is thought to inflict pain on the self. Outrage, in this light, is as much self-directed as it is other directed.

The production of qi has been the subject of many recent studies on activism in contemporary China. When protesters engage in collective action against merciless factory owners or despotic local officials, they often cite "unwillingness to swallow the mouthful of *qi*" as their reason for action.[11] This "mouthful of *qi*" is not amassed in a day but is the result of cumulated experiences of exploitation and repression over a long period of time. Factory workers in China are, more often than not, willing to put up with instances of delayed payments, poor labor conditions, missing safety mechanisms, long hours, and other labor violations. They expect, however, that their patience will add to the relational debt that the factory owner owes them. This accumulated debt, relational or financial, is expected to be paid back by some future acts of beneficence. When they fail to materialize, collective action ensues.[12]

Activists speak of their emotional and behavioral eruptions as a way to "release *qi*."[13] They also frequently refer to factory owners and local officials as people who "lack righteous *qi*."[14] In this context, "righteous *qi*" is an expression of trust and justice in the community. It embodies the commonsense expectation of reciprocity. Isolated instances of repression are written off as outliers, thereby leaving the mass of congenial qi undisturbed. However, recurring and cumulative experience of hardships leads to collective eruption of qi against factory owners or public officials. The community's burst of resentful qi is similar to what sociologist Émile Durkheim termed "collective effervescence," a moment when the torrent of communal sentiment is larger than the sum of individual emotions.[15]

Why are people willing to endure substantial hardships before resorting to collective action? The Chinese idiom, enduring qi and swallowing voice, is often invoked, reflecting a much deeper cultural conviction that qi needs cultivation. A thin-skinned person is described as not having voluminous qi. Every time one endures a challenging moment, qi is thought to mature inside. For example, the *Book of Rites*, the Warring States period (475–221 BC) classic text on social norms and conventions, instructs its readers to "avoid initiating major uproars, in cultivation of qi."[16]

The pursuit of congenial qi figures most prominently in ecological relations. A Tang Dynasty (618–907) folk song, "Golden Well," opens with the words: "Civilization in harmony. Heaven and earth in purity. Qi in unity. And all beings in dignity."[17] Here as elsewhere in classical Chinese literature, qi is thought of as the source of energy behind all life-forms between heaven and earth. For thousands of years, Chinese literati have been inspired by observations of birds and beasts, winds and waters. When these ecological elements recur in Chinese poems, they serve much more than a metaphorical purpose; they appeal to a deep sense of concord between culture and nature.[18] Radiation from the sun breathes qi into all living things. Likewise, plants, animals, and people prosper when they are bountifully filled with earthly qi.[19] This qi-centered worldview prioritizes commonality above everything else. Oneness was the highest ideal, and harmony was the utmost pursuit for ancient Chinese thinkers.

This worldview still lives on in different parts of the Chinese-speaking world, particularly through the practice of feng shui, which literally translates into "wind water" in English. According to the frequently cited Jin Dynasty (265–420) classic, the *Book of Burial*, "Qi disperses with wind and conserves by water."[20] Therefore, the practice of feng shui boils down to a set of principles to conserve qi.[21] Today feng shui is most commonly applied in architecture and interior design, especially in building orientation and interior layout. Feng shui masters specialize in taking advantage of environmental conditions, such as natural light, wind patterns, sloped ground, and local materials, to maximize the qi-conserving potential of the built environment.[22] Auspicious feng shui is believed to emanate from structures that are organically integrated into the surroundings.

While qi is a crucial consideration in spatial design, its upkeep depends on the everyday work of the occupant. Most minimally, the daily conservation of balanced qi is done through opening windows often. In recent years, with high levels of air pollution in major urban centers across Asia, there has been a clash between time-honored practices of open windows and the grim reality of contaminated outdoor air. A team of researchers at Tsinghua University had to warn the public in a recent report that "when the observed value of PM 2.5 exceeds 150 μg/m^3, open-window ventilation becomes a damaging factor for indoor air quality."[23] Such an admonishment, while scientifically sound, stands at odds with the established cultural wisdom. A time-tested cultural preference becomes an ill-advised reference for living through air pollution in the Anthropocene.

But is it? The traditional practice of open windows reflected a willing submission to nature. It was based on a cultural conviction that the natural yin–yang proportion of elements (i.e., in outdoor air) was far superior than that dominated by human influences (i.e., in indoor air).[24] Of course, outdoor air had never been free from human factors—there was the burning, the fermenting, the cooking! But in the outdoors, the human factor was enveloped in a much larger system of ecology. When it came to the quality of the outdoor air, anthropogenic factors were negligible. In this sense, flows of qi from the outdoors helped restore the indoor qi to its primordial state. Open windows served as a connection to nature.

That outdoor qi, however, no longer carries the same natural proportion. Decades of careless growth and development have altered the air, water, earth, and all other ecological elements.[25] It is not that the ancient philosophical construct of qi has gone obsolete but that the commitment to balanced qi has been outrageously violated. In its original use in classical Chinese literature, qi revealed humanity's inherent sense of unity with other earthly beings; it also laid a modest claim to membership in a shared ecosystem.[26] In today's colloquial use, however, *tong qi*, or letting through qi, is reduced to a mere instrumental act of ventilation. Today we keep the form but leave behind the substance.

What's left behind is the ethics of qi: the recognition of qi as cosmic energy that governs social, political, biological, geological, and ecological worlds alike.[27] Such an ethics is not merely a denunciation of modern consumerism or a declaration of deep ecology. It is a near-religious reminder of human inadequacy. In the same way that Chinese naturalist painters intend to meticulously recreate the flow of qi in nature with the flow of ink on paper, living with qi is living consciously under the full spectrum of its ethical principles.

In the epigraph, Father Boyang applied his theory of qi to the analysis of earthquakes. He regarded the three devastating earthquakes as physical proof of the disturbed flow of qi in the dynasty. The yin–yang balance was manifestly in peril, a sure sign of the myriad problems under the king of You's rule, including moral corruption, economic failure, and military vulnerability.[28] Nearly a decade after Father Boyang's prophesy, the king of You swallowed his last breath of qi, as did the dynasty his predecessors had fought to sustain.

Father Boyang provided a holistic diagnosis of the national condition. Rising above the noise of everyday bureaucracy, he was able to focus his attention on the big-picture course of events. He did not treat individual events in isolation but rather prioritized connections between them—qi. If Father Boyang was alive in the Anthropocene, he would probably ask a series of penetrating questions: Would a coal mine nurture the same kind of qi as a solar farm? Would a bicycle bring to its user the same kind of qi as a gas guzzler? Would a fossil fuel–dependent economy foster the same kind of qi as a locally self-sustaining society? What kind of qi does it take for individuals to band together to resist dangerously unsustainable lifestyles?

These questions invite us to think about connections that may not be immediately apparent to the modern mind. They foreground the relationships between the well-being of individuals, communities, and ecological systems. They emphasize that the environmental predicament of the Anthropocene is profoundly connected to various other challenges at the individual and communal levels. In other words, the failure of individuals to overcome narrowly defined notions of self-interest and the failure of communities to act collectively in the common interest are at the root of ecological degradation. The cosmic energy, qi, which powers action at multiple levels, is in disarray. A truly sustainable Anthropocene is only imaginable when a sustainable balance of qi begins to flourish at all levels.

Moreover, qi entails a universal sense of care. The same qi that empowers humans lies in mountains, rivers, landfills, and even parking lots. The notion of qi establishes a complex web of relationships that traverse boundaries, scales, and species. It is indiscriminate.[29]

It should be clear by now that a qi-based worldview departs from the philosophy of the Enlightenment in at least two general fashions. First, instead of treating the individual as the natural unit of analysis, a qi-based worldview situates individual human beings within a network of vibrant life-forms.[30] Rather than elevating humans above all other beings in a food chain–based hierarchy, it throws humans back into the "mosaic ecosystem," to borrow a term from environmental historian William Cronon.[31]

Second, a qi-based worldview makes no distinction between rational interests and emotional sentiments. The flow of qi does not assume that rationality is the only basis of social and individual action. In this conception, individuals make calculated decisions for themselves, but they could be based on rational cost–benefit analyses (e.g., the projected net gain of qi) as well as on sentimental desires (e.g., the urge to release qi). Neither is reducible to the other. A full understanding of social action in the Anthropocene must account for both affective attachments to ecological elements and knowledge of natural degradation.

In these ways qi has the potential to enable a new ecological imagination. Qi is at once personal, communal, and ecological. For ancient

Chinese thinkers, the qi of the individual is sustained through the management of all kinds of worldly desires. It entails sacrifice of hedonism in individual life, endurance of temper in interpersonal relations, and moderation of appetite for materiality. All loom large for life in the Anthropocene.

NOTES

1. Throughout this chapter, when classical Chinese texts are quoted, the author's translations appear in the main text, and the originals appear in the endnote. Original epigraph: 出自《国语·周语》，原文："夫天地之氣，不失其序；若過其序，民亂之也。陽伏而不能出，陰迫而不能蒸，於是有地震。"
2. James Miller, *China's Green Religion: Daoism and the Quest for a Sustainable Future* (New York: Columbia University Press, 2017).
3. Stephen M. Gardiner, *A Perfect Moral Storm: The Ethical Tragedy of Climate Change* (Oxford: Oxford University Press, 2011).
4. David Harvey, *Seventeen Contradictions and the End of Capitalism* (London: Profile Books, 2014).
5. Michael Mayerfeld Bell and Loka L. Ashwood, *An Invitation to Environmental Sociology* (Los Angeles, Calif.: Sage, 2015).
6. Qicheng Zhang, "Qi-Yin/Yang-Five-Elements Model Complexity Re-examined," *China Medial Review* 18, no. 5 (2003): 276–79.
7. Original: 出自《灵枢·五味》，原文："故谷不入，半日則气衰，一日則气少矣。"
8. Guangren Sun, "Yin/Yang Dichotomy in Qi and Qi-Based Theory in TCM," *Journal of Nanjing TCM College (Social Science)* 2, no. 1 (2001): 11–13.
9. Wenwei Chen, "Nature of Qi from Bio-energy Perspective," *Beijing TCM College Review* 17, no. 2 (1994): 7–9.
10. Yali Niu and Weixiong Liang, "Quality of Life Measures with TCM," *Clinical Review China* 10, no. 39 (2006): 144–46.
11. Xing Ying, "Qi and Social Action with Chinese Local Characteristics," *Sociological Research*, no. 5 (2010): 111–29.
12. Yanhua Zeng, "Worker Rebellion from the Perspective of Qi: Case of Factory," *Journal of Anhui Agricultural University (Social Science)* 24, no. 4 (2015): 86–91.
13. Xing Ying, "Qi and Social Action."
14. Kuan Li and Xiaofeng Zhao, "Referenced State: Village Qi and Appeals by Villagers," *Journal of Tianjin Administration Institute*, no. 5 (2015): 59–65.
15. Émile Durkheim, *The Elementary Forms of the Religious Life* (1912; reprint, New York: Free Press, 1965).
16. Original: 出自《禮記·月令》，原文："毋舉大事以搖養氣。"

17. Original: 出自唐朝刘商人《金井歌》，原文："文明化洽天地清，和氣氤氳孕至靈。"

18. Junhan Zhang, "Comparative Analysis of Empathy in Chinese and Western Philosophies," *Journal of Nanjing University*, no. 3 (1996): 59–63, 98.

19. Yu-Ming Liu, "Naturalistic Chi (Qi)-Based Philosophy as a Foundation of Chi (Qi) Theory of Communication," *China Media Research* 4, no. 3 (2008): 83–91.

20. Original: 出自《葬書》，原文："氣乘風則散，界水則止。"

21. Also in the *Book of Burial*, it is said that "ancestors collect [qi] in avoidance of dispersion and maneuvers in pursuit of conservation, hence the notion of feng-shui." Original: 出自《葬書》，原文："古人聚之使不散，行之使有止，故謂之風水。"

22. Lisheng Liu, *Globally Significant China Elements: Folk House* (Taipei: Eculture, 2015).

23. "Tinghua Issues First Big Data Research Report on Indoor PM 2.5 Pollution," Tsinghua University, 2015, http://www.tsinghua.edu.cn/.

24. Yu-Ming Liu, "Wu Ting-Han's Naturalistic Qi-Based Philosophy," *Journal of Religion and Culture of National Cheng Kung University*, no. 5 (2005): 19–58.

25. Bill McKibben, *The End of Nature* (New York: Random House, 1989).

26. Hanglun Zhan, *Propositions in Chinese Literary Aesthetics* (Hong Kong: Hong Kong University Press, 2010).

27. Robert P. Weller, *Discovering Nature: Globalization and Environmental Culture in China and Taiwan* (Cambridge: Cambridge University Press, 2006).

28. Fangmin You, *Primer on Comparative Philosophy between Chinese and Western Traditions* (Taipei: Showwe Information, 2010).

29. Wei-Ming Tu, "The Continuity of Being: Chinese Visions of Nature," in *Nature in Asian Traditions of Thought: Essays in Environmental Philosophy*, ed. J. Baird Callicott and Roger T. Ames (New York: SUNY Press, 1989), 67–78.

30. Jane Bennett, *Vibrant Matter: A Political Ecology of Things* (Durham, N.C.: Duke University Press, 2009).

31. William Cronon, *Changes in the Land: Indians, Colonists, and the Ecology of New England* (New York: Hill & Wang, 1983).

Pronunciation: rén (ˈreːn)

Part of Speech: Noun

Provenance: Chinese, Confucianism, Taoism

Example: "If your ambition is humanity [仁], and if you accomplish humanity [仁], what room is there left for rapacity?"—Confucius, *Analects*, 20.2

Suggested Use: Try to translate rén into a Western language of your choice, and you will inevitably think about your place in the Anthropocene.

The logogram for rén combines two of the most frequently used characters (*hànzì*) in the Chinese language, 人 (human, person) and 二 (two): 仁. In Confucian ethics, rén draws on this etymology and denotes kindness, humaneness, humanity, goodness, authoritative conduct, and benevolence. According to the *Analects* of Confucius (_475–220 BC), rén is the absolute disposition rulers and their subjects should strive for: "Seeking to achieve humanity [仁] leaves no room for evil" (*Analects*, 4.4).[1] Rén, once attained, abolishes all violent social antagonisms. An almost archetypical example of an untranslatable term, Western sinologists, poets, and philosophers have made numerous attempts to translate rén in the many English translations of the *Analects* that have been published in the last century. These translations often provide fascinating insights into Western conceptions of goodness and humaneness, yet the recent emergence of the Anthropocene as a concept in the humanities, as well as the renewed interest in political and ecological virtues it has generated, arguably complicate the evaluation of these translations. In this entry, several classic translations of the *Analects* and of the word rén will serve as a way of thinking about the Anthropocene and about several questionable assumptions the concept involves.

The Confucian rén and the notion of the Anthropocene share a linguistic and conceptual ambiguity. The singular term *anthropos* (ἄνθρωπος, man) at the core of the geological concept of the Anthropocene remains as evasive as the singular 人 (human, person) on which rén is predicated. Indeed, charges of ideological whitewash have been leveled against use of the term "Anthropocene" in the humanities: while the Anthropocene unfolds on a global scale, the *anthropos* (man) at its etymological and causal root may conceal an all-too-familiar *Homo industrialis*. Industrial-capitalist societies, not "mankind" in general, are responsible for the degradation of nature since the beginning of the Industrial Revolution in the late eighteenth century.[2] Can we hold on to the concept of humanity, and hence to humaneness or rén, as a political and ecological virtue, without overlooking causality and culpability?

This question has long captivated readers and translators of Confucius's *Analects*. Generally the Euro-American word *anthropos*, "man," reappears as a kind of cultural contraband in the 人 of most of the early translations of the *Analects*: the way European and American translators have dealt with rén reflects historically and locally constituted discourses. Some of the well-known translators of the *Analects* were poets who had not mastered classical Chinese but were keen to bolster their radical aesthetic aspirations with exoticized translations of Asian literatures. Ezra Pound, for example, plucked the modernist motto "Make It New" out of the *Da Xue* (Great learning), one of the four books of the Confucian tradition. Pound defined rén as "*Humanitas*, humanity, in the full sense of the word, 'manhood.' The man and his full contents."[3] He speaks of the few historical examples of rén incarnate (described in *Analects* 18.1) as "men (with a capital M)," while Canto LXXXIV, in a comment on the same passage, mentions "men full of humanitas (manhood) or jên."

For all their masculinist overtones, Pound's translations manage to convey the importance of education in Confucian ethics. Pound's translation of rén as *humanitas* refers not only to the cultural ideals of Renaissance humanism but also to the fact that the Latin word *humanitas* itself was the way Roman authors liked to translate the ancient Greek word *paideia* (παιδεία), perhaps one of the most complex notions in ancient Greek. *Paideia* is the ancient Greek educational program of "moulding human character in accordance with an ideal," which aimed at making "each individual in the image of [the] community" to which he belonged.[4] Indeed, Pound's use of *humanitas* is suggestive of the Confucian attempt to build community around ideal ethical norms. In the *Analects*, the crucial document of Confucianism attributed to the philosopher Confucius (551–479 BC) and compiled by his followers, the ideal of moral perfection, while being rooted in compassion, is nevertheless a product of ritualized learning (礼, *lǐ*, ritual, ceremony, propriety, convention)—that is, education:

> 3.15 The Master [Confucius] visited the grand temple of the Dynasty. He enquired about everything. Someone said: "Who said this fellow was expert on ritual [礼]? When visiting the grand temple, he had to enquire about everything." Hearing of this, the Master said: "Precisely, this is ritual [礼]."

Ritualized Confucian learning implies a kind of studious docility that was also the essence of ancient Greek education. In Confucianism, but also in Greek *paideia*, moral perfection may only be attained by individuals who are able to renew traditions—after having submitted to them. This, of course, has left the door open for several antidemocratic interpretations of both Confucianism and Greek *paideia* from a Western perspective. Industrial societies consistently make access to cultural resources, and hence to cultural responsibility and political agency, more difficult for the weak and poor than for others. For example, political scientist and ecologist William Ophuls calls for a renewed attention to *paideia* as a means of enforcing the rise of "a natural aristocracy instead of an artificial meritocracy" that could be entrusted with the task of sustainable governance, thus contradicting "the prevailing democratic ethos, according to which there can be no 'betters.'"[5] While the antidemocratic sentiment that transpires here is as old as ancient Greek pedagogy itself, Ophuls's conclusion is somewhat misleading in suggesting that *paideia* (or, by transposition, *humanitas*, and for Pound, 仁) produces a natural hierarchy among citizens. Rather, the political community that supports and enforces the ideals of *paideia* remains aware of the extreme artificiality and contingency of the citizen it manages to "mold" from the clay of the originary *anthropos*. The "beautiful and virtuous" citizen (καλὸς κἀγαθός, *kalos kagathos*) is so precisely because he meets the ideals instituted *by* the community and meets these ideals *for* the community. There can be no question here of a natural aristocracy arising merely from society's perpetuation, especially if this entails this society's ecological self-destruction.[6]

Western conceptions of rén are, then, open to both democratic and antidemocratic interpretations. Considering this ambiguity, and considering the possible value of rén as a virtue, Pound's translation of rén as *humanitas* is both faulty and insightful. In the *Analects*, Confucius suggests that rén "comes from the self, not from anyone else," yet ultimately benefits "the whole world" (12.1). Pound, however, seems to point out the perennial dispute Western civilizations have grappled with since Greek antiquity: can virtue be the object of communal elaboration and transmission? Can virtue be taught?

In raising this question, Pound merely echoes the reaction of several Chinese philosophers to Confucian virtue ethics 2,300 years earlier. Lǎozǐ, the founder of philosophical Taoism and a possible contemporary of Confucius, gives an oblique answer to the question of whether virtue can be taught. Chapter 38 of Lǎozǐ's *Dàodéjīng* (fourth century BC), an "assault"[7] on Confucian virtue ethics, discusses rén in dismissive terms; Brooklyn poet Witter Bynner's (1881–1968) often-quoted pseudo-translation (1944) of the *Dàodéjīng* renders rén [仁] as follows:

> Losing the way of life, men rely first on their fitness;
> Losing fitness, they turn to kindness [仁];
> Losing kindness [仁], they turn to justness;
> Losing justness, they turn to convention [礼].[8]

In Lǎozǐ's *Dàodéjīng*, rén is merely the corrupted form of "the way of life," or *Dao*. Kindness, or rén, stands somewhere between supreme virtue and mere convention, which prompts Lǎozǐ to add: "It is because the most excellent do not strive to excel / That they are of the highest efficacy"[9]—or, in Ursula K. Le Guin's terse and idiosyncratic rendering of the *Dàodéjīng*: "Great power, not clinging to power, / has true power."[10] Le Guin's anarchist sensibility is clearly at play here and brings the profoundly conservative nature of Confucian moralism to light: If rén can be taught, the *Dàodéjīng* wryly suggests, it is precisely as a debased, "teachable" virtue—that is, as a virtue that "clings" to its conceptual delimitations—producing acts that are easily identifiable as virtuous ones (for instance, "greenwashing" the commodities of industrial capitalism), rather than standing as an absolute ethical imperative (for instance, saving the environment from total destruction).[11] Responsibility and humanity, Lǎozǐ and Le Guin suggest, imply a constant alertness to the way words cling to power and facilitate its self-perpetuation.

The various translations of rén discussed here are a case in point, as is Bynner's translation of rén as "kindness." *Kindness* derives from the Neolithic, proto–Indo-European root *ǵénh os* (lineage, race), itself derived from *ǵénh* - (to produce, to give birth). Several further inflections may be observed along the evolution of Indo-European languages: Old English *cynd* (nature, race, family, gender, sex), ancient Greek γένος (race, kind), Latin *genus* (kind, birth, race), Old French *gentil* (noble), and the more contemporary *gentillesse* (kindness). Located at the etymological intersection of nature and politics, of moral dispositions and social order, the word *kindness* eschews attempts at a stable, politically inoffensive definition. Bynner's translation of rén as "kindness" serves as a potent reminder that Indo-European languages strictly predicate "benevolent" moral dispositions and their corresponding social practices on a normative construction of the human—or at the very least they consistently make this predication possible. In other words, Bynner suggests that no language can convey a truly universal conception of benevolence, or indeed kindness. A language always conveys its own social history, and the word *kindness* always speaks the language of a tribe, a clan, race, or a gender.

This is of crucial importance for the examination of the biological and moral notion of humankind, which the concept of the Anthropocene draws on. Wendell Berry quotes Witter Bynner's "losing kindness, they turn to justness" in a discussion of the disintegration of communities and of their essential role "as principle and as fact" for the defense of "ecological, economic, social, and spiritual" health. As an alternative to "litigation,"[12] Berry thinks of "community" as the "indispensable form that can intervene between public and private interests."[13] As we have seen, however, establishing kindness (and, vicariously, 仁) as the moral disposition that is most naturally akin to communal life presents several pitfalls, for the deep semantic layers of *kind* and *kin* could be restored in order to legitimize communities built around notions of race, kinship, class, clan, gender, nobility, or nationhood. Here, rén (as "kindness") might legitimize tribalism and nationalism as responses to climate change and environmental migration.

Yet historians of ideas have long insisted that though common use of the term *kindness* often ignores these tribal, genealogical, and national delimitations, an alternative, universalist construction of kindness may turn out to be equally unsatisfactory. Kindness toward one's universal kin reproduces the same conflation of a moral virtue and a covertly reductive sense of belonging. As Marc Shell remarks, the French revolutionary motto "Liberté, Egalité, Fraternité" and the various religious and political doctrines of universal siblinghood ("All ye are brethren," Matthew 23:8) always imply a logical correlative: "Only my 'brothers' are men, all 'others' are animals and may as well be treated as such," or "Be my brother or I will kill you."[14]

The *Analects* outline a similar tension: "Since a gentleman behaves with reverence and diligence, treating people with deference and courtesy, all within the Four Seas [i.e., ancient China] are his brothers" (12.5). It continues, "Let the lord be a lord; the subject a subject; the father a father; the son a son" (12.11). The clarity of the contradiction between universal brotherhood and patriarchal structures (father and son, lord and subject) seems intentional and enables a central argument of the *Analects* to unfold. Both traditional social structures and the moral imperative of achieving universal siblinghood (which would abrogate these very structures) are declared to be norms in the *Analects*. Confucianism thus holds on to both universalism and particularism, and inserts these norms into its conception of rén, revealing a seemingly unsolvable contradiction. From the perspective of what it seeks to make thinkable or possible, rén points toward a final, universal reconciliation while preserving the social structures on which accountability and responsibility may be instituted. Resolving this contradiction would reconcile what environmentalists fight for with what environmentalists fight with (laws, knowledge, indignation, protest). The Confucian tradition locates rén at the intersection of responsibility and reconciliation, and asserts that reflecting on the tension between these terms is constitutive of superior moral conduct.

Despite the calculated contradiction outlined here, the Confucian tradition has long attempted to clarify the political and ecological implications of instituting rén as a social and moral ideal. Chinese philosopher and Confucian commentator Mencius (~372–289 BC) recounts a parable that proved decisive for the later development of neo-Confucian rationalism: Emperor Xuan witnesses people hauling off an ox to the temple, where its blood is to be drawn to consecrate a new bell. Having seen the ox "shivering with fear like an innocent person," Emperor Xuan intervenes and asks the people to sacrifice a sheep instead. Worried that his gesture might be misinterpreted as an expression of miserliness unbefitting of an emperor, he asks Mencius for advice:

> "No harm done," said Mencius. "That's how Humanity [仁] works. You'd seen the ox, but not the sheep. And when noble-minded people see birds and animals alive, they can't bear to see them die. Hearing them cry out, they can't bear to eat their meat. That's why the noble-minded stay clear of their kitchens."[15]

Cynical as it may appear to be, Mencius's final sentence is not an injunction to "stay clear of the kitchen," to shut one's ears to brutality and exploitation. An indirect reading of this parable is possible, in which Mencius's comments do not only discuss the ambiguities of rén in political practice but circuitously hint at Emperor Xuan's assumption concerning the other moral category at play in this parable: miserliness, meaning Emperor Xuan's fear of being perceived by his subject as ungenerous and selfish. Mencius outlines an ethics of self-restraint and environmental sustainability that suggests that Emperor Xuan's conduct would have been adequate even if it had not stemmed from compassion, hence implying that sacrificing a sheep instead of an ox out of frugality is in itself an act that is both laudable and conductive of rén:

> Mencius said, "The forests were once lovely on Ox Mountain. But as they were near a great city, axes cleared them little by little. Now there's nothing left of their beauty. They rest day

and night, rain and dew falling in plenty, and there's no lack of fresh sprouts. But people graze oxen and sheep there, so the mountain's stripped bare. When people see how bare it is, they think that's all the potential it has. But does this mean this is the nature of ox mountain?

"Without the heart of Humanity [仁] and Duty alive in us, how can we be human? When we abandon this noble heart, it's like cutting those forests: a few axe blows each day, and pretty soon there's nothing left."[16]

Mencius develops and makes somewhat more explicit Confucius's discourse on self-restraint in the *Analects*.[17] The "rapacity" (20.2) of those who do not possess rén is equal to the "covetousness"[18] of those who do not know "when to stop" (1.12). This lack of restraint and sense of shame[19] is illustrated with immediately intelligible poetic images; it also has concrete effects on the biosphere, the degradation of which, Mencius suggests, eventually becomes just as obvious: there can be no such thing as Mencius's kitchen, or any other place where nature may be exploited behind closed doors, in a thoroughly and visibly exhausted environment. Mencius institutes "nature" as a way of thinking about politics in such a way that it remains an extension of virtue—that is, an "extension of ethics."[20]

The sporadic use of rén in recent environmental writing is indicative of the contradictions at the core of the discussions of the Anthropocene currently taking place outside the domain of the natural sciences. At times, and taken as a whole, the canonical translations of rén resemble a sprawling, scholastic inkblot that reveals less of Confucian ethics than it does of the Western longing for a virtue that would meet the two criteria of being apolitical and regulating (in other words, normative); yet these criteria must remain in contradiction with each other if Western societies hope to overcome a politically naive and, more importantly, counterproductive conception of a common "humanity."

In this, rén can be thought of as a conceptual tool to highlight and comprehend several contradictions between Western virtue ethics and the global environmental predicament. The term rén, understood through the lens of its various Euro-American translations, might lead industrial-capitalist societies to reconsider their ambiguous but culturally essential commitment to collective virtues as well as their persistent attempts to deflect responsibility.

As long as industrial-capitalist societies cannot produce a better translation of rén—that is, as long as industrial-capitalist societies need rén as a loanword to think about the conflicting claims of conciliation and responsibility, of virtuous humanity and individual accountability—the most daunting cultural problems posed by the Anthropocene may be considered unsolved. Rén reveals the ethical and political tensions Western societies face in the Anthropocene. Thinking about such a term is one way of addressing them.

NOTES

1. All excerpts from the *Analects* quoted in the text are from the Simon Leys translation unless otherwise noted. Confucius, *The Analects*, ed. Michael Nylan, trans. Simon Leys (New York: Norton, 2014).

2. See Amitav Ghosh, *The Great Derangement: Climate Change and the Unthinkable* (Chicago: University of Chicago Press, 2016), 85–116.

3. Confucius, *Confucius: The Great Digest, The Unwobbling Pivot, The Analects*, trans. Ezra Pound (New York: New Directions, 1969), 22.

4. Werner Jaeger, *Paideia: The Ideals of Greek Culture*, vol. 1, *Archaic Greece: The Mind of Athens* (Oxford: Blackwell, 1946), xxii, xxiv.

5. William Ophuls, *Plato's Revenge: Politics in the Age of Ecology* (Cambridge, Mass.: MIT Press, 2011), 99.

6. See also Bernard Williams, *Shame and Necessity* (Berkeley: University of California Press, 2008), 7–20.

7. *Daodejing: "Making This Life Significant"—A Philosophical Translation*, trans. Roger T. Ames and David L. Hall (New York, Ballantine Books, 2004), 137.

8. Witter Bynner, trans., *The Way of Life According to Laotzu: An American Version* (New York: Perigee, 1994), 67.

9. *Daodejing*, 135.

10. Lao Tzu, *Tao Te Ching: A Book about the Way and the Power of the Way*, trans. Ursula K. Le Guin (Boston: Shambhala, 1998), 52.

11. An argument developed by François Jullien in *De l'Être au Vivre. Lexique euro-chinois de la pensée* (Paris: Gallimard, 2015), 194–98.

12. All quotes are from Wendell Berry, *Sex, Economy, Freedom, and Community* (New York: Pantheon Books, 1993), 139.

13. Berry, *Sex, Economy, Freedom, and Community*, 119.

14. The second quote is attributed to Nicolas Chamfort. Marc Shell, *Children of the Earth: Literature, Politics, and Nationhood* (Oxford: Oxford University Press, 1993), x, 25.

15. *Mencius*, trans. David Hinton (Berkeley, Calif.: Counterpoint, 2015), 27–28.

16. *Mencius*, 148–49.

17. Douglas Robinson further develops this argument in *The Deep Ecology of Rhetoric in Mencius and Aristotle: A Somatic Guide* (Albany: SUNY Press, 2016), 26–48.

18. In his 1938 translation of the *Analects*, Arthur Waley renders 20.2, quoted at the beginning of this entry, as, "If what he longs for and what he gets is Goodness [仁], who can say that he is covetous?" Confucius, *The Analects of Confucius*, trans. Arthur Waley (London: Allen & Unwin, 1945), 233.

19. *Analects*, trans. Leys: "In a country where the Way prevails, it is shameful to remain poor and obscure; in a country which has lost the Way, it is shameful to become rich and honored" (VIII.13).

20. *Analects*, trans. Leys, xvi.

Pronunciation: zen-zookt ('seɪnˌzʊkt)

Part of Speech: Noun

Provenance: German

Example: While touring the ruins of Chernobyl, the sound engineer felt a flash of sehnsucht as her pining for what was lost blended with a joyful comfort in the return of new, if strange, life.

Have you ever had the experience of suddenly recognizing that a familiar, conventional way of solving a problem or performing a task is now obsolete, displaced by something astoundingly new and revolutionary? That mixture of loss and potential gain that you thought and felt was an encounter with sehnsucht.

Sehnsucht, a noun from the German language, dwells within a complex ecosystem of meaning that has evolved over time and space and continues to evolve today. For Friedrich von Schiller, the poet, physician, and philosopher who lived and worked from the mid-eighteenth to the early nineteenth centuries, sehnsucht captured a human experience that was central to German Romanticism and the lives of people in the Western world of his era. For Sigmund Freud, the father of psychoanalysis, sehnsucht helped explain the particular traumas and neuroses of modern life in the first decades of the twentieth century, as people experienced new technoscientific machines such as railway trains. For C. S. Lewis, one of the rare writers to bring the notion into English, sehnsucht conveyed the joyful longing of life as a Christian, looking ever forward to the splendors of Heaven that would follow life on Earth. Novelist Warren Ellis recently joined Lewis in importing sehnsucht to English through his weird and wonderful novel, *Normal* (2016), about a foresight strategist who gets so overwhelmed by the futures he sees approaching that he is sent to recover in a wilderness retreat with other burned-out futurists. This brief list highlights touchstone moments in the history of sehnsucht that we might consider as we shape the loanword's meaning in this time of increasing ecological awareness of mass extinction, toxic pollution, water scarcity, and the Anthropocene epoch writ large, as things we have already lost and the myriad things we will lose once the climatic changes we've set in motion are realized, all alongside the technoscientific breakthroughs and advances that contain the potential to ameliorate these losses.

From those glimpses at the Anglo-Saxon intellectual history of sehnsucht, let's turn to what the word means today. Although sehnsucht is a word on the move through time and space—morphing and adapting as people utter it within shifting conditions of life on Earth and across diverse cultural settings—its two root components remain *sehn* ("to long" or "to pine") and *sucht* (a lingering illness). A consistent if somewhat reductive definition combines two elements. One is pining, as in a longing for something lost, a something that may have once existed and is now gone, or a something that may have been an opportunity passed over or a potential never realized. The other is anticipation, as in a deep, perhaps utopian desire for a better state of affairs to come, though this future may reside anywhere on the scale that runs from the realistically realizable to a state of perfection that can only be approached asymptotically. Such a combination of sadness and joy, of intellect and feeling, seems to describe a paradox: to encounter sehnsucht is to oscillate between despair and hope, loss and gain, dystopia and utopia.

In other words, sehnsucht captures a complicated, one might

say dialectical, response to living in a world that feels like it's on the precipice of disintegrating and/or nearing the horizon of a radical breakthrough into a new and humane future—a synchronization of unflinching acceptance and utopian hopefulness. It's like the transient vision human beings get at dusk when the photoreceptors of their retinas toggle uncertainly. We may feel as if we can't see what we expect to see clearly, yet new colors and contours come into view. This paradoxical experience of being is not an easy space to inhabit for any length of time, really, so sehnsucht is necessarily a fleeting and precious thinking-feeling.

By conscientiously integrating sehnsucht into thought, art, science, and general discussions of past, present, and future ecological crises, people stand to improve their capacity to overcome the debilitating despair that often flows from awareness of these crises. To face mass extinctions, the deteriorating conditions of water, air, and soil on Earth, and the massive-scale hyperobjects of global warming and the Anthropocene with sehnsucht is to acknowledge devastation without turning in desperation to nostalgic pining for some sort of idealized past world of ecological harmony and balance that never in fact existed.[1] Nor can we turn to fantasy projections of humane capitalism that can keep things running essentially as they are now, but with smart phones made with bamboo cases. No. Sehnsucht helps us to look unflinchingly at the horrors of life on Earth, now and to come, and maintain hope for and enjoyment in the radically alternate futures that cannot yet be designed even as it feels like such futures are reaching out to us, transmitting their coordinates so we can birth them into being in a strange and retroactive time loop. Sehnsucht is a pleasure in the ruins of today that might help propel us toward the unknown pleasures on the far side of ecological horrors; sehnsucht might sustain the will to work through grim times toward the prospect of brighter futures.

In order to bring sehnsucht into current ecological conversations, it is useful to examine the aesthetic powers it wielded in the past to articulate human perceptions of being but a part in a massive and complex world. One of the most powerful uses of this loanword in German is the 1801 poem "Sehnsucht" by Friedrich Schiller.[2] Published at a time when sehnsucht possessed significant cultural capital in the German cultural milieu, Schiller's text circulated widely. The poem was so popular that it inspired renowned musical creations by Franz Schubert and Siegfried Wagner. Just as the term sehnsucht is a complex intersection of multiple intellectual and emotional valences, so too was Schiller a brilliant polymath—poet, physician, philosopher, historian. It's hard not to wax nostalgic for a social context in which forceful expressions about human coexistence with the nonhuman elements of the planet emanated from both scientists and aestheticists. To esteem the arts, humanities, and sciences as equal, perhaps mutually imbricated, modes of epistemology—if still in dynamic tension with each other—is an integrated approach to problem solving that is sorely needed today as we confront Anthropocenic crises. Yet the idea of Nature that emerged during the Romantic era and solidified Nature into a stable thing that became a commodity and/or externality haunts the present. In short, Nature facilitates human enlightenment, and enlightened consciousness is what gives Nature the meaning that it would otherwise not possess.

Schiller's poem gorgeously embodies the paradoxical Romantic-era sehnsucht experience of deeply embracing the nonhuman world while maintaining an image of the human being as an individual, separate and seemingly autonomous from those nonhuman elements of Earth. The first two and a half stanzas of the four-stanza poem present readers with a multisensory description of wandering outside. The reader is prompted to feel "chilling mists," glimpse "hilled dominions, / Young and green eternally," hear "Dulcet concords" and "sweet celestial calm," smell on the breeze "sweetly fragrant balm," and imagine encountering a full-body refreshment from inhaling "the air in highlands yonder." By activating all of the senses, Schiller invites readers into the poetic

equivalent of a virtual reality immersion into this awe-inspiring location. The poem "Sehnsucht" was built to capture that fleeting but forceful experience of walking outdoors in synchronicity with the nonhuman
world. But, like a technical glitch in a virtual reality rig, this immersion
dissolves when the person in the poem comes across a raging river and
reacts by perceiving a gap—a deep separation. Good news, though, as
the poem takes a tonal turn: there is a boat, and the protagonist decides
to launch it, powered by faith and daring. The poem ends with a declaration: "Only a wonder can you carry / To the lovely wonderland."
Schiller's "Sehnsucht" offers a hopeful vision of mustering the will to
cross the impasse that separates us from the wonderland of the full
world. But the catch is that this shooting the gulf—to borrow a phrase
that Ralph Waldo Emerson used in "Self-Reliance" (1841) to argue that
power resides not in repose but in transition—maintains and further
empowers the image of the human will to dominate Nature. Even if the
intentions portrayed in scenes like Schiller's poem and Emerson's essay
prompt readers not to rest in ideas that can ossify, these same intentions often seem to entail controlling and manipulating nonhuman elements of the world for our purpose and benefit.

In 1930, in the wake of the horrors of World War I, in the midst
of the rapid advancement of Hitler's rising power and the Great
Depression, Sigmund Freud published one of his most lauded works,
Civilization and Its Discontents. Sehnsucht appears fairly early in this
book-length analysis of why people in the rapidly modernizing areas
of Europe felt such deep malaise and alienation despite the astounding
rise of technoscientific developments. Freud writes:

> During the last few generations mankind has made an
> extraordinary advance in the natural sciences and in their
> technical application and has established his control over
> nature in a way never before imagined. . . . Men are proud

of these achievements, and have a right to be. But they seem
to have observed that this newly-won power over space and
time, this subjugation of the forces of nature, which is the
fulfillment of a longing [*Sehnsucht*] that goes back thousands
of years, has not increased the amount of pleasurable
satisfaction which they may expect from life and has not
made them feel happier.[3]

Sehnsucht is key to this passage, as Freud suggests that this human
relationship to the nonhuman world is a desire that is very likely prehistoric. By implication, then, it seems unsurprising that the inventions of
the steam engine, railroad, and cinema, for example, would fully satisfy
this desire. To be human is to encounter sehnsucht; to be human is
to pine for synchronicity with the nonhuman elements of Earth and
feel disalienated from it, and paradoxically to achieve control over those
same elements of Earth, so that we feel masters of it. In other words,
Freud deployed sehnsucht to identify the cognitive dissonance he perceived as a widespread human response to the lived experience of being
part of a planet's ecosystems. But the human part of this formulation
and what it means for sehnsucht as a loanword is complicated. After
all, Freud was analyzing lives in largely Western urban environments
placed in conversation with lives in often rural or agricultural non-
Western environments. He theorized with sehnsucht in a fuzzy zone
where cultural specificity commingles with human universality, which
works out well for us because sehnsucht responds to a universal condition based on a shared planet, yet it comes marked by cultural specifici-
ties that do not disavow the differences in how climate futures will be
experienced by human beings.

Intriguingly, Freud claims that the cognitive dissonance of sehnsucht has intensified because of the simultaneous explosion of technoscientific developments in the West and contact through global
expeditions with what Freud, in the parlance of his time, refers to as

"primitive peoples and races."[4] In a weird way, Freud's sehnsucht resonates with the affects of H.G. Wells's *War of the Worlds* (1897), a work
of science fiction that combines the allegory of inverted invasion—
Martians treat British people the way British people were treating
Tasmanian people—with the prospects of advanced technoscience to
reshape an entire planet. Indeed, science fiction written in the era of
European colonial expansion expresses the time's hubris and anxiety.
It remains a genre that captures as well as reimagines the sehnsucht
state of being.

Now, in the early decades of the twenty-first century, people live
in the gloom of looming global warming, haunted by the prospect of
human life flickering out on Earth. Structures and practices of globalization such as the existing and potential uses of algorithms, bioge-
netics, and geoengineering differ from those that Freud witnessed, yet
they are still marked by the contradictions of that era and fears that
such advances require sacrificing essential aspects of what some people
think it means to be human. Sehnsucht can play a significant role in
the present moment, but it must be wielded wisely to be effective. If
the pining side of sehnsucht is overemphasized, it may fuel retrograde
ideas of making Earth great again, to modify a Trumpism; if the hopes
of terraforming other planets or fixing Earth with nanomachines are
overemphasized, sehnsucht may lead us into tempting fantasies of
technoutopianism; but by holding and grappling with both the painful acceptance of the past and the hopeful speculations of the future, a
dynamic sehnsucht can help articulate the deeply unsettling experience
of recognizing the world in and as transition. To feel-think sehnsucht
right now is to comprehend and even embrace how alien life on Earth
has become without ever forgetting that the human species is inextricably bound to this planet through millennia of natural selection. Sehnsucht is an awful and awesome revelation that any single human being
seems able to sustain only briefly, so as to avoid being pulled apart by
its forceful contradiction.

It's as if we're a somnambulist species who's been remodeling unconsciously for millennia, only to wake up one morning and discover new rooms and hallways, plumbing and wiring and heating ducts all around us. We are at home, yet not. We feel sehnsucht—longing for a better world that we've never inhabited but that we believe to our cores exists as an alternative. It must! So, please consider making sehnsucht a word you can shake out of your sleeve when you catch those fleeting glimpses of paradox, of pivot points—when you grasp the temporary ruins around us, not to shore them up, but to reimagine a whole new life of joys and pains, of arts and sciences, of coexistences and loves right here on Earth as we all warm up together into the future, foreseeable and beyond.

NOTES

1. Timothy Morton, *Hyperobjects: Philosophy and Ecology after the End of the World* (Minneapolis: University of Minnesota Press, 2013).
2. Friedrich Schiller, "Longing," *Poems and Ballads by Friedrich Schiller* (1759–1805), Schiller Institute, http://www.schillerinstitute.org/.
3. Sigmund Freud, *Civilization and Its Discontents*, trans. James Strachey (New York: Norton, 1961), 39.
4. Freud, *Civilization*, 38.

Ghurba
ANOTHER PATH

Pronunciation: shi-cata ga-nai

(çi̥kata ga nai)

Part of Speech: Colloquialism

Provenance: Japanese and speculative fiction
(Kim Stanley Robinson's Mars trilogy)

Example: To those who say that transforming our
petroculture is too big a task, we say shikata ga nai.

Shikata ga nai is a historically layered term that speaks to the delicate
balance between pragmatism and idealism—a critical negotiation as we
face an ecologically compromised future. A Japanese-language phrase,
shikata ga nai translates roughly as "it cannot be helped."[1] During
and after World War II, it was commonly used largely by issei—first-
generation Japanese American and Japanese Canadian immigrants.
Policies of racial profiling and confinement were undergirded by similar attitudes on either side of the U.S.–Canadian border, resulting in the
displacement of entire communities, from infants to the elderly, without due process. During the war, people were forced into internment
camps or to work; at war's end, they faced resettlement to mandated
zones of North America or "repatriation" to Japan.[2] During this period,
shikata ga nai expressed a collective resignation to these policies, which
produced incalculable suffering. It also signaled an acceptance by the
speakers that these policies were an unfortunate but perhaps inevitable outcome of World War II. By uttering shikata ga nai and other-

wise remaining silent about this experience of injustice, many Japanese Americans and Japanese Canadians endeavored to assimilate and move on.[3] Most were successful.

Yet in spite of the resignation that shikata ga nai expressed, we see within it a kernel of collective possibility, a glimpse of how best to work toward a positive future in difficult times. The generation that came of age in the postwar period, composed largely of nisei and some sansei (whose parents and grandparents, respectively, were born in Japan), was frustrated by their treatment during and after the war. Inspired by the civil rights movement, they sought to discover their silenced histories, reverse the attitude of quiet acceptance, and fight for official redress. Their struggle succeeded in gaining formal apologies from both the Canadian and American governments. In 1988, each acknowledged that other choices were possible, even during wartime. Suddenly what was inevitable began to shift.

The wartime history of shikata ga nai plays an important role in Kim Stanley Robinson's Mars trilogy (comprising *Red Mars* [1993], *Green Mars* [1994], and *Blue Mars* [1996]), one of the major works of Western science fiction in the last three decades. Written in the shadow of the 1988 formal redress settlements, the trilogy shows how the phrase might be productively recontextualized for environmental politics without erasing its racialized history.[4] Our proposed adoption of this loanword looks to the lived experiences of Japanese Americans and Japanese Canadians rebuilding their lives in the ruins of racist World War II government policies, on the one hand, and an imagined fictional future where humans are forging new communities on Mars in partial response to environmental catastrophe on Earth, on the other.[5] In this context, shikata ga nai as loanword reflects a need to persevere despite traumatic and even shameful histories, drawing a line between historical policies that were intended to erase racial minorities and contemporary environmental crimes.[6] Shikata ga nai asks all of us to recognize that in the face of the seemingly insurmountable, we have to be resolute and willing to redefine what is inevitable. In the current and coming moments of climate crisis, this lesson will offer some shelter to weather the storm. As per our suggested use to those who say that transforming our petroculture is too big a task, we say shikata ga nai. As the redress movement accomplished in the preceding decade, Robinson's saga about the future colonization of Mars transforms the meaning of shikata ga nai from passive acceptance of historical injustices and hardships to a politically expedient mobilization. In the Mars trilogy, shikata ga nai signals the sense that the politics of the inevitable can be flipped; in the case of the present, shikata ga nai is a bridge built toward a necessary and welcome political and ecological transformation.

Robinson's Mars trilogy captures the malleability of shikata ga nai: characters mobilize a logic of inevitability and practicality to justify decisions about the future of Mars. These visions are in competition with one another: will Mars be terraformed or not? If so, how? At stake in these decisions is not only the ecology of a historically lifeless planet but also the question of the eventual, real-world geoengineering of planet Earth—once science fiction but now a very real possibility. In *Red Mars*, the first book in the series, there are at least four different visions for the future of Mars, led by four women who arrive as part of the First Hundred (botanists, a psychologist, scientists, and engineers): Hiroko Ai, Nadia Cherneshevsky, Ann Clayborne, and Phyllis Boyle. Robinson fashions these four female characters as a political synecdoche. Each embodies one of several attitudes about the relationship of Earth colonists to Mars. Shikata ga nai as a disposition is associated with three of these leaders as they work to shape the future. Although their plans for Mars are quite different, three of the four involve extracting resources from and manipulating their new environment to meet the needs of human communities. In each instance, the phrase downplays the ecopolitical significance of their decisions.

Ai first introduces shikata ga nai in *Red Mars* during a debate over which energy sources to rely on for the development of Martian settlements. Born in Japan, Aï's story-world history could easily be entwined with the real Japanese Americans and Japanese Canadians who used the phrase, especially if one reads the trilogy as an extension of Earth's real history. They debate the merits of nuclear power capacity and wind power infrastructure, with the American Phyllis Boyle in the corner of capitalist expedience and the Russian Arkady Bogdanov arguing for a radical new vision of collective social life:

> It was Hiroko who cut Arkady off with what she said was
> a Japanese commonplace: "*Shikata ga nai*," meaning *there
> is no other choice*. Windmills might have generated enough
> power, but they didn't have windmills, while they had been
> supplied with a Rickover nuclear reactor, built by the U.S.
> Navy and a beautiful piece of work; and no one wanted to try
> bootstrapping themselves into a wind-powered system, they
> were in too much of a hurry. *Shikata ga nai*. This too became
> one of their oft-repeated phrases.[7]

Though she stops the argument, Ai has ulterior motives: the nuclear power generators will work better to serve her goal of forming a hidden colony to develop a truly Martian culture. To quicken the realization of her secret goal, Ai declares "there is no other choice" than to opt for nuclear power solutions. Though not capitalistic in motive, this expediency echoes economic justifications that rely on the privileging of short-term gains over long-term ramifications, a mind-set that has resulted in disastrous climate change here on Earth.[8] To declare that "there is no other viable option" attempts to evacuate this critical energy infrastructure decision (to implement nuclear power on Mars) of its politics. Furthermore, the absence of a shared vision for the future of Mars, and an energy strategy for achieving these visions, means the First Hundred use capitalist productivity quotients (value ÷ time) as a justification for acting in their own interests, sometimes driven by differing ecopolitical visions and sometimes seemingly motivated by scientific curiosity and a desire to exercise their technological know-how, looking to implement the most efficient or practical short-term, if not long-term, solution.

In the trilogy, both Russian and American characters grab hold of Ai's term, shikata ga nai. Nadia Cherneshevsky, the Russian cold-weather engineer, thinks through the practical problem of having enough water on Mars:

> They needed more water, but the seismic scans were finding
> no evidence of ice aquifers underground, and Ann thought
> there weren't any in the region. They had to continue to
> rely on the air miners, or scrape up regolith [Martian soil]
> and load it into the soil–water distilleries. But Nadia didn't
> like to overwork the distilleries, because they had been
> manufactured by a French–Hungarian–Chinese consortium,
> and were sure to wear out if used for bulk work.
> But that was life on Mars, it was a dry place. *Shikata
> ga nai.*[9]

While the colonists eventually develop methods to introduce more water into Mars's planetary system, Cherneshevsky's use of the phrase during the initial settlement establishes pragmatism as the governing ideology. Here, shikata ga nai means "it cannot be helped." Rather than acknowledging the political and ecological implications of her multiple decisions about infrastructure, Cherneshevsky addresses the practical, mechanical issues in front of her. She accepts the phrase shikata ga nai not for its political expediency but because it allows her to continue with her work. There is no need to acknowledge the larger forces shaping the direction of development on Mars (such as Earth-based metanational corporations) so long as the projects succeed and the colonists have enough air, water, sustenance, and shelter. For Cherneshevsky, shikata ga nai names a form of acceptance in the face of what, from her vantage, are merely engineering problems, but which are ultimately political in that they fulfill the official mandate of the mission: to colonize and prepare Mars for the arrival of waves of immigrants.

Cherneshevsky faces immediate crises that demand solutions, while the American Boyle marches toward a grand vision of Martian resource development. As Boyle contemplates the energy well of Martian gravity, she takes on the attitude of capitalistic pragmatism.[10] Her monumental plan: to build a space elevator. In a conversation with another American, John Boone, she says,

> "This gets us out of our gravity well, eliminating it as a physical and economic problem. That's crucial; without that we'll be bypassed, we'll be like Australia in the nineteenth century, too far away to be a significant part of the world economy. People will pass us by and mine the asteroid directly, because the asteroid has physical wealth without gravitational constraints. Without the elevator we could become a backwater."
>
> *Shikata ga nai*, John thought sardonically.[11]

The "it" in shikata ga nai makes space for ambiguity, and here Boone's cynicism manifests as a refusal to engage politically. "It" can be read as either resignation to the fact that Boyle cannot resist her capitalist impulses or as ambivalence about Mars becoming a backwater. Boone, as an observer of Boyle's planning, amid a range of other political actors, thinks *"shikata ga nai"* ironically, because he can see quite clearly that there are alternatives beyond succumbing to the economic imperative or being rendered obsolete. This use of the phrase might describe how many people currently experience resignation to a vision of the doomed fate of the Earth. Yet this ambiguity also presents us with an opportunity to argue for a shift in what "it" is that cannot be helped: from a seemingly inevitable future of capitalism and ecological devastation to a viable and resolute response to the ongoing global environmental crisis.

Finally, Ann Clayborne takes an explicit position against development, extractivism, and scientific progress for its own sake. From the outset of the trilogy, Robinson highlights the misguided belief that Mars exists only through human intervention, much as European colonizers believed their gaze made manifest the New World. The dominant impulse to treat Mars as a resource triggers Clayborne's dogged opposition to terraforming the planet. *Red Mars* begins the trilogy with the line, "Mars was empty before we came."[12] From the arrival of the First Hundred, two ways of seeing Mars emerge: those interested in terraforming (called, ironically, the Greens) and those who see it as a place with its own integrity, despite its surplus of rocks (the Reds). Clayborne's desires reach beyond the conflicts of the others, and it is for this reason that she does not utter shikata ga nai. In spite of her disinterest of entering into politics, people treat her attitude as a rallying point, crafting a revolutionary movement that has a major impact on the eventual Martian constitution. Clayborne's refusal to admit "it cannot be helped" wins the Reds several concessions, such as the guarantee that certain areas of the Red Planet will not be touched by the terraforming efforts. Though Clayborne is less than thrilled at these compromises, the results of her passionate engagement complete the picture of a struggle over the future of Mars. Clayborne critiques the underlying assumptions made by those saying shikata ga nai. She flips the referent, inverting the (eco)logic and the worldview of what "cannot be helped." Instead of assuming that "it" is terraforming, what becomes inevitable for Clayborne is the need to leave some of the planet as they found it. Shikata ga nai.

From its first use in the Mars trilogy, the logic of shikata ga nai resonates with its historical usage—as a means of efficiently and productively forging forward by remaining silent about the unjust politics of detainment, dispossession, and dispersal experienced by the Japanese

Americans and Japanese Canadians. Ultimately the lesson of the post-war era was that these injustices could not be forever suppressed, let alone redressed, through resignation; resistance and a robust politics of collective action was needed to drive the redress movement for almost two decades until the governments apologized. Since then, scholars, writers, filmmakers, and artists have been rewriting the official histories from the perspective of Japanese Americans and Japanese Canadians. In borrowing the phrase shikata ga nai, Robinson subtly invokes not only histories of injustice but also the notion that whatever decision is made, even those born of necessity, there may in the future be need for redress and reconciliation.

This need arises today in the realm of environmental policy as many prioritize the maintenance of "the economy." It is taken as a given that one cannot act differently, and that the only commonsense approach to the future lies in creating policy that balances the economy and the environment. But common sense is a dangerous measure for any problem requiring radical departure from current ways of being and doing. Policy makers around the globe seem unwilling to disrupt the accumulation of capital or make decisions against corporate interests. Rather than deep and immediate decarbonization, government decision makers are influenced by corporate lobbyists to ensure that the energy transition is slowed or delayed indefinitely. It's now clear that the decision to press forward with business as usual will not stand the test of time; those who have allowed for these delays in the interest of the corporate wealth will be judged for the devastating "slow violence" that failing to address climate change is already causing.[13] As in the redress movement, shikata ga nai could be a part of an environmental politics that seeks to foreground multispecies ecological justice in decision making.

We do not imagine the comparison between the real-world histories of Japanese North Americans and an imagined future history to be simple or easy. Yet shikata ga nai can contain disparate histories of use in a doubly resonant meaning—one that features a reverberation of the past into the future. Just as people in the past suffered the loss of home and lands, displacement and restrictions on mobility, so too will millions of climate refugees in the coming years. The "it" in "it cannot be helped" is the fact that our ways of doing and being in the world— our relation to one another, other species, and the environment—must change. The "it" that "cannot be helped" is that although we are deeply enmeshed and complicit in the causes of climate change, even though we lack an objective and clear answer about how best to respond, we must act. There will be many experiments and many failures, but each action toward addressing climate change and every failed experiment will yield necessary lessons.

As a loanword, shikata ga nai reminds us that accepting that "it cannot be helped" has constituted an act of survival specific to many times and places. To ensure a livable planet for humans and nonhumans alike, we need to use this loanword to disrupt capitalist relations, colonial relations, and patriarchal relations. Robinson argues for these goals in the Mars trilogy through the polyvocality of the story line that provides a mirror onto the real-world politics of climate change now as it did in the 1990s. Whereas Robinson illustrates how shikata ga nai is a malleable signifier that can be used to obscure political end goals, we argue that its use could—in the real-world politics of the current ecological crisis—be mobilized as a valuable loanword. On this planet, in the twenty-first century, we can confidently say shikata ga nai: there is no other way forward than through radical environmental politics and policies.

NOTES

We acknowledge the support of Jesse Goldstein, Sha LaBare, Natalie Loveless, Matthew Schneider-Mayerson, and Terri Tomsky, who provided input and feedback.

1. Roy Miki, *Redress: Inside the Japanese Canadian Call for Justice* (Vancouver: Raincoast Books, 2004), 260; and Mitsuye Yamada, "'You Should Not Be Invisible':

An Interview with Mitsuye Yamada," conducted by Caroline Kyungah Hong, Shirley Geok-Lin Lim, and Sharon Tang-Quon, *Contemporary Women's Writing* 8, no. 1 (2014): 15.

2. The euphemism of "repatriation" used at the end of World War II overwrites the fact that Japanese Canadians, born in Canada, were among those sent back to Japan—a country they had never been to. This only adds to a list of injustices and their legacies, which are too numerous to articulate here. For more details, see Miki, *Redress*, and Sheena Wilson, "*Obāchan's Garden*: Maternal Genealogies as Resistance in Canadian Experimental Documentary," in *Screening Motherhood in Contemporary World Cinema*, ed. Asma Sayed (Bradford, Ont.: Demeter Press, 2016).

3. Assimilation involved more than remaining silent about the historical civil and human rights abuses they had to endure. It meant denying their language and culture as well. As Roy Miki writes, "Many Japanese Canadians did not have the language to account for the unspeakable monstrosities that manifested themselves internally as shame and guilt for being singled out, ostracized and labeled the 'enemy alien' within the social body of their own country," and therefore people reverted to "common phrases . . . to mediate a past that refused resolution" such as shikata ga nai. Miki, *Redress*, 260. See also Lucia Lorenzi, "Shikata Ga Nai: Mapping Japanese Canadian Melancholy in the Field of National and Literary Trauma," *West Coast Line* 71 (2011): 100–105.

4. In 1988 both the U.S. and Canadian governments offered apologizes to Japanese Americans and Japanese Canadians and began their respective processes of financially compensating former internees. Nineteen ninety-three, the year *Red Mars* was released, was the year that marked the completion of the distribution of the financial settlement in the United States.

5. In mid-February 1942, President Roosevelt signed Executive Order 9066, which lead to the removal of 110,000 Japanese Americans from the West Coast. Less than a week later, the Canadian government passed Order-in-Council 1486 that allowed for the removal of Japanese Canadians from the West Coast, given that a previous order had declared it a protected area.

6. Muriel Kitagawa captures this sensibility well when she says, "Who knows but that the next time will be made easier for the plunderers because we shrugged and said: 'Shikata ga nai.'" Kitagawa, *This Is My Own: Letters to Wes and Other Writings on Japanese Canadians, 1941–1948*, ed. Roy Miki (Vancouver: Talon Books, 2010), 216.

7. Kim Stanley Robinson, *Red Mars* (New York: Bantam Books, 1993), 109.

8. As opposed to the fictitious geoengineering on Mars, and then, of course, the experiments and proposals to use terraforming on Earth in the twenty-first century—

well after the trilogy was written—as a response to manage anthropogenic climate change.

9. Robinson, *Red Mars*, 124.

10. This metaphor describes the gravitational limit to spaceflight. Humans look out to the stars as if from the bottom of a well. Only with the proper tool (a jetpack or even a ladder) could they hope to climb out.

11. Robinson, *Red Mars*, 307.

12. Robinson, *Red Mars*, 2.

13. See Rob Nixon, *Slow Violence and the Environmentalism of the Poor* (Cambridge, Mass.: Harvard University Press, 2013).

Sila

Janet Tamalik McGrath

Pronunciation: see-lah (siːlɐ)

Provenance: Inuktut

Part of Speech: Noun

Example: We are never more than sila.

In the language of Inuit, Indigenous people of the polar Arctic regions, the word sila to refers to many interconnected concepts, depending on context: outdoors, outer, globe, Earth, atmosphere, weather, air, sky, intellect, intelligence, spirit, energy, cosmos, space, universe, and even life force.[1] If adopted as a loanword into English, it would be used to indicate the interconnection of all phenomena, with an understanding that humans have a role in learning how to live in balance with their inner and outer environments. Currently the magnitude of negative human impacts on the Earth's biosphere reflect a disconnection from nature and a false perception of superiority. Using the loanword sila, human responsibility might be expressed in the phrase "We are never more than sila." This implies that with our abilities to impact nature in many ways, we also bear serious responsibilities that need to be acknowledged. What we do to our natural environment and to other beings, we do to ourselves; we have an obligation to endeavor to live in harmony with all forms of life.

The cosmic–human connection is built into the Inuktut language. For example, the root word for wisdom, *silatujuq*, literally translates as "has a lot of sila" and is used when natural and cosmological environments are manifested within a person. This term also refers to clever-ness and cunning—that is, having the ability to see strategically based on inner attributes and thinking resourcefully with a sense of one's responsibilities to act ethically. If one is foolish, ignorant, or lacking in common sense, if one is unaware of one's negative impact on themselves and others, they might be called *silaittuq*, "without sila."

Sila's life force is inside of all humans through its sacred movement—our breath. At the same time it is the source of all breath and life. This connection to our breath implies our integral interconnect-edness with all of creation; wherever energy flows, it also flows within us. This multifaceted sila is not directed by humans, but humans are subject to its gifts, forces, and powers. Consequently, the Inuit elders I work with have emphasized the need to relate to sila with humility, appreciation, and great sensitivity.[2]

Inuktut is quite old. The language in recognizable form dates back 5,000 years, with some clear links to Uralic, which puts it up to 10,000 years old in origin.[3] The word and its significance remained in the realm of oral tradition until Inuk author Rachel Qitsualik contended in 1998 that sila is a "super-concept, both immanent and transcendent in scope . . . arguably the most important concept in classic Inuit thought" because it is simultaneously "intellectual, biological, psychological, environmental, locational and geographical."[4] Sila's immanence is also based on physical reality; no creature can breathe or exist without air, and space is required for material manifestation. Its transcendence is illustrated in its daily expression in Inuktut, as diverse phenomena found within and beyond human realms are related. Because of the importance of weather patterns to Inuit survival and these intercon-necting realms of sila, Qitsualik argued that it reflects both human fragility and responsibility. Humans rely on sila, which supports food sources. Death gives rise to life. We can only live if plants and animals die; this truism is a point of sacred reverence in Inuit culture.[5] Qitsua-lik used the term *superconcept* to convey how sila links together many interrelated concepts, referring to so many processes, beliefs, and aspects of life that it defies any concise or straightforward English translation. To appreciate the worldview accompanying the applications of the term sila, it is necessary to understand some basic features of Inuktut language.

INUKTUT LANGUAGE: AN INTERCONNECTED WORLDVIEW

Inuktut language is structured to convey Inuit ways of knowing, which include the core Inuit values of cooperation, harmony, observation, respect, innovation, and resourcefulness. These values are based on a worldview that is fundamentally relational. One feature of the language that illustrates the importance of relationality is the complex system of pronoun suffixes—transitive verb agreement endings—that include both subject and object as one unit. Though all Inuktut dialects use this form, I draw on examples from Canadian Inuktut. These Inuktut verb endings keep an implicit focus on the importance of relationality, as this key and distinguishing feature of the language is extraordinarily precise about subject–object integrated reality. There are dozens of these fused verb endings, which variously cross-reference the participants in the action. Further, there is no gender or animate/inanimate distinction in Inuktut, and thus the attention of the listener is necessarily focused on context to determine what is actually meant (whether he, she, or it). By affirming connection and obscuring difference, Inuktut brings a speaker's attention to relationality. Some examples include *takujara,* "I see him/her/it." The verb is *taku-,* "to see," with the ending *-jara,* "I to him/her/it" (I→him/her/it).[6] In the English versions of this sentence (I see him/I see her/I see it), "I" is the subject, the seer, and "him/her/it" is the object, the seen. However, the Inuktut version unifies the subject–object relation and even asserts a lack of division, while obscuring gender and the animate/inanimate distinction. These other common endings illustrate how complex Inuktut can seem in comparison to English or other modern languages:

-jait, you/it
   (*takujait* = you see him/her/it)

-jatit, you/they
   (*takujatit* = you see them)

-janga, he/it
   (*takujanga* = he/she/it sees him/her/it)

-jaanga, it/I
   (*takujaanga* = he/she/it sees me)

-jaatigut, it/us
   (*takujaatigut* = he/she/it sees us)

Regardless of the length of a sentence, the morphemes (units of meaning) include verb action and interconnected relation. These verb agreement endings have many variations to convey other moods such as dubitative, conditional, causative, and interrogative. So, for example:

takungmangaagu
   (whether he/she/it sees you him/her/it)

takukpagu
   (when/if he/she/it sees him/her/it)

takungmagu
   (because he/she/it sees him/her/it)

takuvauk?
   (does he/she/it see him/her/it?)

Inuktut is economical in that a speaker changes the word's ending instead of adding another word. As a result, one needs to know many versions of verb agreement endings in order to communicate at even a basic level.

   A challenge that Inuktut presents for an English speaker is the mirror-image sequence, compared to English, of meaningful units (i.e., morphemes) in a single phrase. The subject is identified only at the end of a long phrase, so Inuktut can seem ambiguous to a foreign speaker: everything about the action is listed before who is taking the action becomes evident. With this kind of structure, the participants of the action (the subject and the object) are put in a place of relative insignificance, as the action, activity, time, and space are foregrounded through this basic sequencing (morphology and order indicated in parentheses):

-tunga (I)

Niuvirvingmunngauniaqtunga
(the store-going to-will-I): I am going to go to the store.

-tugut (we 3+)

Niuvirvingmunngauniaqtugut
(the store-going to-will-we 3+): We are going to go
to the store.

-tut (they 3+)

Niuvirvingmunngauniaqtut
(the store-going to-will-they 3+): They are going
to go to the store.

In Inuktut, the emphasis is thus not on the subject but on the subject within a vast range of relevant elements, as in the examples above— place, space, time (store, motion toward, future tense). Inuktut acknowledges that humans are dependent on place, space, and time, and our doer-ness is tacked on the end. In the hierarchy of thought and the flow of communication, we literally come at the end, consistent with our lesser significance. A listener therefore needs to be patient until the actor is revealed. Consider the following sentence: "*Tuktuturumajualutuinnauniraqtauqattaqpallailijualuulauqsimagaluarama.*" This sentence/

word contains eighteen morphemes and translates loosely into English as "Because coincidentally I'd perhaps been referred to by others as only wanting to continuously eat caribou meat." But in Inuktut, the communication begins with *tuktu* (caribou) and ends with *-rama* (because I), with all the other details in between. The "I" subject of this sentence in Inuktut (in its causative form, *-rama*) is only revealed after sixty-four letters, or seventeen other morphemes.[7] In fact, longer examples than the one above are possible while remaining grammatically correct and intelligible. In this way, Inuktut narrative forms naturally cultivate deep listening, presence of mind, and a practice of patience.

With these kinds of complexities embedded in the very structure of the language, is it appropriate to propose that an Inuktut term be adopted into the English language? Absolutely. It is not necessary to understand Inuktut grammar or speech in order to connect to the rich heritage of a single term, and the Inuktut language, with its philosophical underpinnings, offers simple ways to express relationality and interconnection. And sila is only a four-letter word! Additionally, there are already a few commonly used loanwords from Inuktut, including such well-known words as igloo (*iglu*, "snow house") or kayak (*qajaq*, a uniquely Inuit invention, a slim, covered one- or two-person boat). While these are material technologies that reflect Inuit practical innovation, sila as a concept offers a deeper philosophical connection that is critical at this time in world history. Igloo and kayak relate to survival in materially bare circumstances; sila is another kind of survival, and as a loanword in English, it relates to material, emotional, spiritual, social, and environmental survival in a globalized world.

SILA IN ENGLISH

How does one apply such a term that refers to an interconnectedness of all forms and forces, that, while including humans, is not human centered? Here I offer some suggestions, with the expectation that once incorporated into English, other usages will certainly be found. In practical terms, when the weather is contrary to human wishes or needs, one can say "it's sila" to affirm that one needs to work with what is presented, that we are wiser for being with what is rather than complaining and wishing it was otherwise. The experience of awe for nature, as when sun, wind, water, sky, and Earth all meet the senses in a moment of joy and connection, can be expressed in the affirmation, "Sila!" We are at once connected with this beauty, an integral part of it, and indebted to it. And "It's sila!" would be appropriate for the indescribable blessing of knowing we are part of the beauty and integrity of creation, an emotional connection to human life. The word sila encompasses experiences of awe or joy while also affirming a fundamental connection with the forces that inspire these feelings. What is outside (environmental) is inside too. We are interconnected.

Not all of sila's expressions are positive from a human perspective. Natural disasters cause fear and destruction. Sila could offer an enriched view of these phenomena. "By sila . . ." would refer to a rebalancing of forces beyond human control. This would include any natural disasters—storms, cyclones, monsoons, or earthquakes. Though undeniably unfortunate, instead of taking a hostile view of these processes, we might express how these forces are rebalancing phenomena by acknowledging sila. With this acknowledgment, issues stemming from anthropogenic climate change and extreme weather patterns are not denied but become the basis for a renewed resolve to live in balance.

"For sila . . ." would refer to a human action that can benefit the integrity and balance of the inner and outer realms, honoring the interconnection between humans and the other-than-human world. By adopting a "for sila" approach, the cosmic–human balance could be seen as beneficial on many levels, beyond human existence. Additionally, proenvironmental high-impact practices such as living free of cars, avoiding airplane travel, eating a plant-based diet, and reducing family size,[8] combined with planting trees, conserving electricity, using less heat and air-conditioning, and recycling are practical actions that have economical and environmental benefits.[9] However, no term exists in English to express the emotion of being committed to living in harmony with nature. By saying it is "for sila," we can affirm our interconnectedness with nature, express concern about our impacts, and convey our commitment to learning and adapting.

Responding to the challenges of the Anthropocene will involve developing a culture of sustainability. Transmitting better attitudes and practices to future generations is critical. Raising children to have a strong bond with nature now requires a disciplined effort, as most adults' and children's hours are spent within homes and buildings—offices, workplaces, schools—sheltered from natural forces.[10] The term sila would be useful as a means of engaging children—for example, asking them, "What's sila doing?" This would be similar to saying, "Let's observe the weather," but it involves direct sensation—going outside and sensing the air, slight gusts of wind, light breezes, cloud movement, humidity, sun and moon positions and states, animal presence, and changes in plant cycles, thereby affirming the interconnected nature of reality. Using the term sila offers a connotation of connection that the English words *nature* and *environment* lack. Sila offers children a new expression that links external physical reality and emotional being. "Let's spend the day with sila" would be similar to a planned day in nature, except it explicitly acknowledges that the basis of life is a loving bond with all the natural forces of which we humans are a part. If it rains, for example, rather than an opportunity for complaint or a sprint for shelter, it's simply sila feeding the plants. All forces of sila are explored in their support of diverse forms of life and seasonal and geological changes.

What might it mean for English speakers to consider the grammar of putting themselves at the end of a sequence of thoughts? What might it mean to describe everything else first—the weather, other beings, histories of place, and then finally attend to the subjects of an action? Sila is a word that has been lived by for centuries in the polar Arctic. As climate change impacts the Arctic at a rate more than twice the rest of the globe, and as the fast-melting ice impacts us all globally, I offer this word as a bridge between an Inuktut worldview and modern English.[11]

By adopting the Inuktut term sila into English, a timely and vital connection to ancient human values can be reestablished in our current setting. It can enhance an understanding of our position as humans within our Earth's complex systems and beyond, as it reminds us of our responsibility to learn to live in balance.

NOTES

1. Emilie Cameron, Rebecca Mearns, and Janet Tamalik McGrath, "Translating Climate Change: Adaptation, Resilience, and Climate Politics in Nunavut, Canada," in *Annals of the Association of American Geographers* (London: Taylor & Francis, 2015). See also Rachel Qitsualik, "Inummarik: Self-Soverignty in Classic Inuit Thought," in *Nilliajut: Inuit Perspectives on Security, Patriotism and Sovereignty* (Ottawa: Inuit Qaujisarvingat, 2013), 29. Expression and usage may vary by dialect. In some dialects it is pronounced "hila."

2. It is important in Inuit culture to name elders who teach you specific customs, values, and ideas, and not use a blanket statement of authority such as "the elders say . . ." In this case this teaching comes from Mariano Aupilarjuk and Nilaulaaq Aglukkaq, though their teachings are echoed in a wide variety of oral history literature. See John Bennett and Susan Rowley, eds., *Uqalurait: An Oral History of Nunavut* (Montreal: McGill-Queens University Press, 2004).

3. Louis-Jacques Dorais, *The Language of Inuit: Syntax, Semantics, and Society in the Arctic* (Montreal: McGill-Queens University Press, 2010), 101. There are single word resemblances in modern Asiatic languages that suggest Inuktut is even older than the Uralic connection. For example, the word sila bears similarities to the term for outside, space, and sky, *sula* in Manchu, and to the Japanese word for the same, *sora*. Dorais, *Language of Inuit*, 92.

4. Qitsualik, "Inummarik," 29; see also Rachel Attituq Qitsualik, "Word and Will—Part Two: Words and the Substance of Life," special to *Nunatsiaq News*, November 12, 1998.

5. See Qitsualik, "Inummarik." Also note that even though this is a hunting cultural perspective, even a non–meat eater still relies on the death of plants to live, and the same reverence to one's source of life would apply.

6. The pronouns used here reflect the normative, binary genders of English and not the nongendered, animate–inanimate inclusive form of Inuktut. In order to convey the temporality of Inuktut, the pronouns he and she are used here. This is in no way meant to frame gender as a fixed binary, which it is not. It is critical to convey that, though changing, English lacks what Inuktut abounds in, which only furthers the argument made here for sila as an ecotopian loanword.

7. For those interested in the morphological breakdown, each forward slash indicates a separate unit of meaning (dropped consonants and vowels in square brackets): *tuktu/tu[q]/ruma/ju[q]/[a]alu/tuinna[q]/u/niraq/tau/qattaq/pallaili/ju[q]/[a]alu/u/lauq/sima/galua[q]/rama.*

8. See Seth Wynes and Kimberly A. Nicholas, "The Climate Mitigation Gap: Education and Government Recommendations Miss the Most Effective Individual Actions," *Environmental Research Letters* 12, no. 7 (2017): 1–9.

9. These recommendations are for the masses of first world countries based on scientific analyses. Because a loanword from Inuit language is being proposed, it needs to be clarified that in much of Inuit homelands globally, airplane is the only means of travel. Also, from the perspective of first world administrators, many third world economic and health markers are experienced by Inuit. It is also counterproductive to impose a plant-based diet where plants do not grow and where wild animals are still hunted as an essential source of food, not to mention the environmental costs of shipping food. Note also the scale of impact, where for example in Nunavut there is a population density of only 0.02 people per square kilometer. Census Canada denotes this as 0.0, compared to Ontario, which has is 14.1 people per square kilometer.

10. Richard Louv, *Last Child in the Woods: Saving our Children from Nature-Deficit Disorder* (Chapel Hill, N.C.: Algonquin Books of Chapel Hill, 2008).

11. See Sheila Watt-Cloutier, *The Right to Be Cold: One Woman's Story of Protecting Her Culture, the Arctic and the Whole Planet* (Toronto: Penguin Canada, 2015).

Solastalgia

Kimberly Skye Richards

Pronunciation: sah-luh-stal-juh (sɒləsˈtældʒ(i)ə)

Part of Speech: Noun

Provenance: Ecopsychology

Example: Since the mine began clearing woodlands and wildlife corridors, many residents of the area have been suffering from stress, sleep deprivation, anxiety, and solastalgia.

The film begins with an attractive blonde woman stripped of her clothes by an excavator operator on a game show. The television audience raucously applauds as the male operator skillfully maneuvers the steel machinery to clench its teeth on her red silk slip and dramatically undress the woman. The pornographic exposure of the woman's body in the studio setting draws attention to the perverted ways in which human bodies are stripped bare by the machinery of late capitalism. The shot fades to black before picturing an extreme close-up of rashes covering the face of a different woman as an Italian language instructor suggests that listeners learn how to ask for help when confronting the pain, distress, and melancholia that comes from the destruction of one's homeland. Jenny Brown's 2016 film *The Hitchhiker's Guide to the Symbiocene* grapples with solastalgia, the feeling of distress caused by negatively perceived changes to the environment. Solastalgia is described by the language instructor as "gut-wrenching mental anguish because a bulldozer is destroying my beloved streetscape right now." It

266

is the pain felt due to the inability to return to a loved place because it is irrevocably changed for the worse. The instructor's statements will resonate with readers of this volume: "I'm feeling powerlessness in the face of multinational corporations and authoritarian government," and "I can't direct my grief about needing environmental change towards anything or anyone in particular." But the woman with the blistered face does not appear to have the words to describe her psychological and physical anguish. The linguistic lesson and the woman's despair draw attention to the need for a new vocabulary to express the forms of grief and mourning that are emerging in the Anthropocene.

The term solastalgia was developed by Glenn Albrecht, a conservationist and environmental philosopher who was inspired by the people of the Upper Hunter region of New South Wales, Australia, a site of open-cut coal mining, pollution, and drought. Albrecht observed that residents of the Upper Hunter region seemed to be suffering from the sick landscape, and coined the term solastalgia to describe the sense of powerlessness and grief experienced by people when their homeland is under duress. Solastalgia draws on the Latin word *solacium* (comfort) and the Greek root *algia* (pain, suffering, sickness), conveying the anxiety caused by the inability to derive solace from one's home in the face of distressing events. It is part and parcel of a new abnormal of the Anthropocene, characterized by uncertainty, unpredictability, chaos, relentless change, and deep distress caused by a changing climate, erratic weather, and species extinction.[1] Solastalgia might be precipitated by the dwindling numbers of salmon in a river; the eradication of buffalo on the plains; the hyperextraction of natural resources through logging, mining, and tar sands development; or urbanization, through the construction of condos, ski hills, and golf courses. As Albrecht summarizes, solastalgia is "the pain experienced when there is recognition that the place where one resides and that one loves is under immediate assault. . . . [It] is a form of home sickness one gets when one is still at 'home.'"[2]

In the early 2000s, Albrecht identified the need to develop a language to describe how human psychic and somatic health is affected by changes to local ecosystems. He created a typology of "psychoterratic states": earth-related (terratic) mental (psychic) conditions that draw attention to the mental effects of exponentially expanding human development and climatic change.[3] Psychoterratic pathologies illustrate that a sense of place, and the resulting sense of belonging, are crucial to mental health and psychological well-being. Solastalgia is derived from nostalgia, the melancholia or distress experienced by individuals separated from their home, or the wistful desire to return to a former time or place. Coined by Johannes Hofer, a Swiss doctor at the end of the seventeenth century, nostalgia was frequently associated with soldiers fighting in foreign lands who experienced such severe homesickness that they were unable to perform their duties. It was initially considered to be a medically diagnosable illness, and doctors recommended that the afflicted return home to restore their health. Although the concept of nostalgia has taken on new significance in the context of late capitalism and is now more often associated with a desire to return to a romanticized period in the past, solastalgia relates to the sense of nostalgia as a diagnosable illness associated with homesickness for one's native land. Solastalgia also grows out of the word *topophilia*, a positive earth emotion coined by the poet W. H. Auden in 1947 to laud the rich descriptions of beloved places in the poetry of John Betjeman.[4] Topophilia describes the rich attachments and affective bonds we have to the environment.

Topophilia is especially intense for those who live and work closely with the land and who draw cultural, political, psychic, or spiritual sustenance from it. Indigenous people in particular are likely to suffer from environmental changes. Their solastalgia can result in what Colin Tatz describes as "existential suicide"—the ennui, hopelessness, and lack of motivation to live that contributes to the tragically high rates of Indigenous suicide in many communities.[5] Solastalgia is also experienced by farming families suffering from a new wave of aggressive colonization—what geographer David Harvey has called the "new imperialism"—by extractive industries.[6] Solastalgia might be said to describe the experience of Inuit communities in northern Canada faced with rising Arctic temperatures; the inhabitants of the vanishing Pacific Islands; mostly black, working-class residents of New Orleans after Hurricane Katrina; Sioux water protectors at Standing Rock; the Ogoni and Ijaw peoples in the Niger Delta; and other victims of settler colonial conquests that degrade and damage the ecosystem. Previously there was no word in the English language that encapsulated the nature of this suffering so acutely.

In order to make something of our solastalgia, we need to acknowledge it. Solastalgia offers this ecotopian lexicon a term that does not idealize the ecological relations of the past but describes one of the increasingly common socio-ecological experiences occurring today. Although a global condition, solastalgia has spread unevenly around the world and within nations. It is not a new feeling. It emerges from the increasingly dystopian world of the Anthropocene and describes a response to the endangerment, extinction, habitat loss, enclosure, and privatization that have been felt in the third world for some time now, and which Indigenous people have experienced in their homelands since contact. Neshnabé (Potawatomi) scholar-activist Kyle Powys Whyte explains that we already inhabit what our Indigenous ancestors would have understood as a dystopian future.[7] Albrecht also identifies a range of well-known poets and artists who were expressing the experience of solastalgia before it had a name, including Romanticists such as William Wordsworth, who represented the gradual loss of a loved home environment; Salvador Dalí's portrayals of the desolation of mind and landscape; and Edvard Munch's *The Scream* (1893), depicting a contorted figure in anguish set against a lurid red sky.[8] These well-known poems and works of art convey the feelings of disorientation, homelessness, anxiety, depression, grief, despair, estrangement, and powerlessness that result from the slow destruction of communities, neighborhoods, sacred places, and ecosystems. They speak to how our lives are entangled in ecological, cultural, and economic relationships; they also speak to the difficulty of learning from, adapting to, and responding to the challenges we face today.

While the experience of solastalgia is increasingly familiar, what remains to be determined is the path of healing. What, if anything, might relieve the pain and anguish of the loss of one's beloved home? In 2015, Jenny Brown curated an event called "Solastalgia" at the Cementa15 Arts Festival in Kandos, New South Wales. Brown was inspired by a scene in Ray Bradbury's dystopian novel *Fahrenheit 451* (1953), in which members of the resistance attempt to commit banned books to memory in order to preserve them. Brown cast members of the community as "book custodians" and directed them to read excerpts from the writings of German political theorist Hannah Arendt, French sociologist Pierre Bourdieu, Australian nature writer Sharon Munro, American poet-philosopher Henry David Thoreau, and the local newspaper. At the same time, scenes from Francois Truffaut's film version of *Fahrenheit 451* (1966) and interviews of community members were projected on the train's doorway and windows. While environmental and political philosophers are not censored as they are in Bradbury's book, Brown's performance suggests there is therapeutic value in communal events in which environmental knowledge can be shared for those afflicted with solastalgia. In publicly acknowledging the deterioration of the biosphere, individuals become less alienated in their bereavement, and might collectively develop strategies for resistance.

Although solastalgia and other psychoterratic ailments are challenging to treat because victims are often unable to direct their grief and anger toward anything or anyone in particular, Brown's performance concluded with the audience being asked to set alight a pile of yule logs named after Australia's then prime minister, Tony Abbott, representing the settler colonial regime that enforces the laws and policies of extractivist coal mining in the name of economic development. These pyrotechnics created a cleansing pagan ritual for the melancholy of solastalgia and other negative effects of extractivism. As art critic Ann Finegan notes, Brown's participatory performance offered the distressed community consolation and "an occasion for a small rural village to demonstrate the effects of globalized mal-development to the broader urban arts community."[9] Beyond the art world, solastalgia indicates the value of naming psychoterratic states and providing a space for communities to gather despite the shrinking commons, to share knowledge about protecting, restoring, and rehabilitating their homelands, and to help individuals overcome the dread, alienation, and disempowerment of contemporary politics. The feeling of solastalgia does not immediately motivate citizens to change their consumptive, personal, or political practices, but it does run counter to climate change denial and avoidance, and its acknowledgment can galvanize productive political projects.

Love for the land can inspire a politics of love. In an interview about the Indigenous environmental movement Idle No More with Canadian author and activist Naomi Klein, Mississauga Nishnaabeg scholar Leanne Betasamosake Simpson articulates the importance of developing an intimate and loving relationship with the land even if it has been depleted, degraded, or damaged. She remarks, "When I think of the land as my mother or if I think of it as a familial relation, I don't hate my mother because she's sick, or because she's been abused. I don't stop visiting her because she's been in an abusive relationship and she has scars and bruises. If anything, you need to intensify that relationship because it's a relationship of nurturing and caring."[10] Despite the damage caused by settler societies and extractive industries, there is still so much to love about nature. Solastalgia is indicative of an estrangement from a place because of its transformation, but not all is lost, and places must not be abandoned because they are not perfect or pretty. Developing a loving relationship with a place despite the sadness and trauma of witnessing its destruction is an important act of resilience in the Anthropocene, and one that holds a tremendous healing power.

NOTES

1. Glenn Albrecht, "Exiting the Anthropocene and Entering the Symbiocene," Glenn Albrecht (blog), December 17, 2017, https://glennaalbrecht.wordpress.com/.

2. Glenn Albrecht, "'Solastalgia': A New Concept in Health and Identity," *PAN: Philosophy, Activism, Nature* 3 (2005): 48.

3. Glenn Albrecht, "Solastalgia and the Creation of New Ways of Living," in *Nature and Culture: Rebuilding Lost Connections*, ed. Sarah Pilgrim and Jules N. Pretty (London: Earthscan, 2010), 218.

4. W. H. Auden, introduction to *Slick but Not Streamlined*, by John Betjeman (New York: Doubleday, 1947). See also Yi-Fu Tuan, *Topophilia: A Study of Environmental Perception, Attitudes, and Values* (New York: Columbia University Press, 1974).

5. Quoted in Albrecht, "Solastalgia," 51. The concept of existential suicide is based on Albert Camus's philosophy about of the meaninglessness of life, ennui, and lack of motivation to exist. See Colin Tatz, *Aboriginal Suicide Is Different: A Portrait of Life and Self-Destruction* (Canberra: Aboriginal Studies Press, 2001).

6. David Harvey, *The New Imperialism* (Oxford: Oxford University Press, 2003).

7. See Kyle Powys Whyte, "Our Ancestors' Dystopia Now: Indigenous Conservation and the Anthropocene," in *Routledge Companion to the Environmental Humanities*, ed. Ursula Heise, Jon Christensen, and Michelle Niemann (New York: Routledge, 2017).

8. Glenn Albrecht, "Solastalgia and Art," *Mammut Magazine* 4 (2010): 12.

9. Ann Finegan, "Solastalgia and Its Cure," *Artlink* 36, no. 3 (2016), https://www.artlink.com.au/.

10. Leanne Betasamosake Simpson, "Dancing the World into Being: A Conversation with Idle No More's Leanne Simpson," conducted by Naomi Klein, *Yes! Magazine*, December 5, 2013, https://www.yesmagazine.org/.

Pronunciation: swé·ño ('swe.ɲo)

Part of Speech: Noun

Provenance: Spanish

Example: Life is un sueño.

Oil-soaked, coal-roasted, gasoline-delivered: the American Dream epitomizes fossil-fueled existence and promises a nightmare in the ongoing planetary ecological crisis. The American Dream associates the word *dream* with the Great Acceleration—the bundle of human pressures (fossil fuel use, population growth, mass urbanization, deforestation) that, rapidly increasing around 1950, pushed the Earth into its Anthropocene configuration.[1] The consequences are now emerging: global warming, sea-level rise, extreme weather, mass extinction, societal breakdowns, refugee crises. Given such trends and the Great Acceleration's continuation into the early twenty-first century, Earth system scientists wonder whether humans will decouple civilization from fossil fuels in time to dodge irreversible ecological catastrophe: "Will the next 50 years bring the Great Decoupling or the Great Collapse?"[2] They suggest that by 2050, a century after the Great Acceleration took off, "We'll almost certainly know the answer."[3]

The likely inventor of the phrase "American Dream," historian James Truslow Adams, defined the dream in *The Epic of America* (1931): "That American dream of a better, richer, and happier life for all our citizens of every rank, which is the greatest contribution we have made to the thought and welfare of the world."[4] When Adams exalted the American Dream during the Depression, while millions struggled with economic devastation, the phrase named a shared hope. A couple of decades later, the economic growth characterizing the Great Acceleration imprinted on individual and collective psyches a lasting expectation of the American Dream's inevitable fulfillment. After World War II, the United States surfed the crest of a global growth wave far exceeding any previously seen in the industrial period: "In the half century after 1950, the global economy grew sixfold," with growth "peak[ing] between 1950 and 1973."[5] Although unevenly distributed globally and even within the United States, the benefits were felt most strongly among Americans, an apparent fulfillment of the American Dream.

As viral as blue jeans, this dream found its defining realization in the 1950s. In the United States from 1947 to 1960, "GNP per capita increased 24 percent" and "personal consumption spending increased by 22 percent."[6] "In the 1950s," amid the postwar baby boom, "with only 7 percent of the global population, the United States . . . accounted for half of global manufacturing output and accrued nearly half of the world's income each year."[7] The Time-Life Books series Our American Century includes The American Dream: The 1950s, and a 1959 Saturday Evening Post cover pictures the Dream: beneath the moon, a young white couple rests against a tree with the stars above constellated into a home, a garage, "a pool, two cars, two pets, three children, a stereo, a television, a washer and dryer, a drill press, [and] an air conditioner."[8] Most Americans alive today were born under that constellation. Of the current U.S. resident population (approximately 327 million as of July 2017), about 84 percent were born in 1953 or later.[9] As historians J. R. McNeill and Peter Engelke note, a clear majority of humans now alive have spent their entire existence within the Great Acceleration, "the most anomalous and unrepresentative period in the 200,000-year-long history of relations between our species and the biosphere."[10] The perceptions, desires, and memories of most living Americans are confined to the era of the Great Acceleration.

An artifact of an unprecedented historical moment, the 1950s American Dream influenced subsequent history by shaping its dreamers' consciousness. How does this occur? Philosopher Bernard Stiegler notes that sensations, memories, and dreams intertwine.[11] Following the phenomenologist Edmund Husserl, Stiegler calls sensations "primary retentions." Memories are past sensations worked into "secondary retentions." For Husserl, individuals apprehend objects via memory's selective editing of sensation. Memory's edits influence apprehension—perception itself. But for Stiegler, collective memories always inform this process. Consider two Americans, both born after 1945, encountering a specific car. The American who possesses memories of that car apprehends it differently than the one who does not. But before ever driving, both had inherited collective dreams about cars, the "signature commodity" of the 1950s American Dream.[12] Stiegler calls inherited dreams, such as the dream of home ownership, "tertiary retentions." Tertiary retentions format the interplay of primary and secondary retentions, of sensations and memories. Neither the term "myth" nor the phrase "cultural heritage" quite defines tertiary retentions, which are mnemotechnical: they are collective memories that media technologies convey across generations. Technologies capable of handing down collective memories include print (novels, histories, leaflets), film (newsreels, documentaries, propaganda shorts), and digital new media (smartphones, tablets, laptops). Americans now google images of dream homes and cars; these pixelated images mediate their perceptions and memories of the cars they drive and the houses they occupy.

For a movie that hands down a peculiarly American collective memory of the Great Acceleration, consider 1968's *Bullitt* (dir. Peter Yates).[13] In 1968, a boy sits in a theater watching actor Steve McQueen as police sergeant Frank Bullitt careens his Ford Mustang down city streets to pursue villains fleeing in a Dodge Charger, which finally crashes into a gas station (cue high-octane fireball). This famous chase sequence traversing the San Francisco hills constitutes a tertiary retention of the Great Acceleration. Besides advertising fossil-fueled vehicles via film, a medium inseparable from oil,[14] the sequence encodes a collective memory of the rush of power, speed, and dominance that the Great Acceleration afforded select Americans. A human-initiated planetary event, the Great Acceleration became manifest to a specific population through sensations coded as "I'm in the driver's seat." *Bullitt* retains a collective memory of those sensations so coded. Never having driven, without personal memories of driving, the boy in the theater dreams about driving. His dream implicitly anticipates his future as a continuation of the American Dream's transient, unsustainable, fossil-fueled realization. The carbon dioxide from the cars he will drive lingers in our atmosphere still.

With the Great Acceleration leaving the biosphere awash in the Anthropocene, to continue dreaming the fossil-fueled American Dream results in films such as *Fast and Furious 8* (2017) and in the presidential administration of Donald J. Trump (January 20, 2017–?). Dreaming the future as an irrecoverable past's triumphant return ("Make America great again"), the Trump administration is (as of this writing) accelerating Anthropocene processes with dystopian potential. This acceleration ties the American Dream to a definition of U.S. sovereignty as somehow unacceptably compromised by, and incompatible with, international climate treaties, pushing the dream further into climate change denial. Proponents of expanding fossil fuel extraction—for example, by opening Alaska's Arctic National Wildlife Refuge to oil drilling—must turn the dream from hope for the future toward a violent nostalgia for the past. American English needs a loanword that would help to open the language to ecotopian possibilities—a future of solar-only cities, post-carbon hedonism, cross-species solidarity, and so on—by reengineering our dreams, American or other, to welcome sovereignty's delimitation, climate change science, and genuine hope for the future.

Consider the Spanish noun sueño as that loanword. Sueño refers both to sleep and to the dreams that occur in sleep. *Tengo sueño* translates as "I am sleepy," while "I have a dream" may be rendered as *tengo un sueño*. This essay focuses on sueño as dream, though sueño as sleep plays an important role too.

Addressing the climate crisis necessitates renegotiating sovereignty. Around the resurgent question of sovereignty, renowned seventeenth-century Spanish dramatist Pedro Calderón de la Barca may help Americans (and others) to dream differently. In un sueño, borders between day and night, waking and sleeping, and existence and inexistence become permeable. The wavering of borders un sueño may trigger occurs when Calderón stages a fathomless, unnerving convergence of living and dreaming in his play *La Vida es Sueño* (Life is a Dream) (1636). Calderón's play opens after Basilio, Poland's king, has imprisoned his infant son, Segismundo, in an isolated tower because of an astrological prophecy that his son would become a tyrannical monarch bent on crushing his father. Basilio would defy the stars. After his imprisoned son comes of age, Basilio tests the prophecy by having minions drug Segismundo and bring him to the palace, where Segismundo awakens to find himself king. Should Segismundo prove a viable sovereign, the stars will have been refuted. But when Segismundo turns despotic (to prove his will boundless, he throws a servant out a high window) and so seems to confirm the prophecy, Basilio drugs him again and returns him to the tower. Upon awakening, he is told that his brief life as king was merely un sueño. With narcotics and subterfuge, Basilio attempts to replicate and stage-manage the labyrinthine interlacing of life and dream that is the play's larger theme. This dynamic exceeds and undoes Basilio's attempt to conjure a future amenable to his sovereignty. Returned to prison, his father's stratagem ironically leads Segismundo to accept the interlacing of life and dream—that duplicitous simulacrum conjured by

Basilio. Attempting to master fate by seeking to foil the prophecy that his son will abase him, Basilio stages kingship as un sueño. At the play's end, after the populace turns against Basilio and frees Segismundo to name him king, the wisdom of his first acts stems, he says, from his having let un sueño be his "maestro."[15] Consider un sueño as potentially both sovereignty's remedy and its poison. Calderón stages life's inter-lacing with dream as simultaneously the strategy of the sovereign and the sovereign's ultimate undoing. Although Calderón's vision is irreduc-ible to the American Dream, the latter tends to congeal into a dream of fossil-fueled sovereignty—dreamt by 1950s-era United States as the oil-guzzling global hegemon, but also dreamt by Americans for whom being at the wheel signifies the will's autonomy. Somewhat like a self-driving electric car with an MP3 of Dr. Seuss's *Oh, the Places You'll Go!* playing on continuous loop, *La Vida es Sueño* ironizes sovereignty as a dream that entrances the sovereign. Rather than signaling identification with or participation in sovereignty, in Calderón's play, encountering *la vida* as un sueño reveals sovereignty as evanescent and contingent, ripe for the undoing.

To counter many American dreamers' increasing aversion to knowledge and science, consider Calderón's avid reader, Sor Juana Inés de la Cruz, a poet of seventeenth-century New Spain so famous in lit-erary tradition that she was known as the tenth muse. In her poem "El sueño," Sor Juana describes the sueño realm as a liberation from daytime conventions and authorities.[16] Early in her life, before becom-ing a nun, Sor Juana secured viceregal patronage, which was crucial to the network of allies that allowed her to pursue her literary ambi-tions. Sor Juana's "El Sueño" elaborates a dream vision in which the soul takes flight, surveying the world's wonders and soaring skyward to seek the pinnacle of knowledge. Nighttime becomes the soul's ally. The poem begins with night falling and sleep overtaking all, including those inclined to police Sor Juana during the day, from the calculating suitors who sought Sor Juana's hand in marriage during her time at the viceregal court to the clerics who sought her silence after she became a nun. In Sor Juana's sueño, borders of time and space give way. The soul envisions the famous lighthouse of Alexandria and an even more ancient Egyptian wonder, the pyramids. At the poem's close, with daylight's return, the soul's flight ends as the sleeper wakes. Sor Juana's poem narrates a dream that opens a realm of free inquiry and liberates the soul so the speaker's desire for knowledge finds limits more in the cognitive capacities of the human than in the restraints of censorious authority. Climate scientists in the contemporary United States might envy Sor Juana her sueño. She would empathize with their resistance to ongoing efforts to defund and ignore climate science.

As for Calderón, so for Sor Juana. When authorities seek to imprison the mind or to silence thought—the Trump administration's efforts to shut down research on global warming, for example—the time has arrived to embrace un sueño, for in un sueño desire and hope may flourish and serve as motivation for resistance. Consider the Spanish civil war. During an October 12, 1936, public gathering at the University of Salamanca, a region held by the fascists, philosopher Miguel de Unamuno confronted the fascist officer and propagandist José Millán-Astray. With the fascists in attendance chanting "Viva la muerte" (Long live death), Millán-Astray shouted at Unamuno, "Muera la inteligencia" (both "Death to the intellect!" and "Death to the intellectuals!").[17] The next day, Unamuno was expelled from the university.

But even amid fascism's ascendancy in civil war–era Spain, un sueño manifested, at least to Mexican essayist, poet, and public intellectual Octavio Paz. Recounting his experiences of the Spanish civil war in *El Laberinto de la Soledad* (The labyrinth of solitude), Paz describes how he saw in the faces of the antifascist fighters un sueño made flesh: their faces shone with incarnate utopian hope, "la Esperanza."[18] Paz recounts how this vision sparked within him an unquenchable desire to once again encounter human beings profoundly transformed by "la Esperanza," a luminous hope shining in an utterly bleak time. The sueño Paz encountered in waking life exemplified and prefigured, if only fleetingly, a "reconciliación del hombre con el universo," a reconciliation of humanity with the universe.[19] The utopia Paz hopes for contains environmental resonance, as this utopia features harmony between the human and the nonhuman—in Paz's terms, "el hombre y la naturaleza."[20]

How would sueño becoming a loanword in English help to activate the ecotopian imagination? In the United States, the word sueño would alter English speakers' collective dreams—that is, their collective memories—pushing those dreams to become less stereotypical. Stereotypes whiten out the object triggering them, whomever or whatever that object may be. The more a collective memory involves the stereotypical, the more that collective memory works on individuals' sensations and memories to airbrush a given object from consciousness, as when stereotyping leads to invisibility. Stalin ordered comrades he disfavored erased from photos to scrub their existence from collective memories and to deny reality. Alternative facts, fake news, and climate change denial build on the stereotype potential of collective dreams. Yet collective dreams may also work on sensations and memories so as to underscore the reality of an object's existence. Political struggles often entail reimagining collective dreams to bring the reality of given humans (or nonhumans) to consciousness. Collective dreams that sharpen an object's reality for consciousness are what French philosopher Bernard Stiegler terms traumatypical. While stereotypes reinforce conservative impulses, traumatypes might help rearrange our collective dreaming. In contemporary Anthropocene circumstances, the ecotopian imagination welcomes the way that traumatypes underscore an object's reality, while climate denialism allies with the object's disappearance into stereotypes. That a collective memory's traumatype potential may contribute to the ecotopian finds support from novelist Iris Murdoch: "Love is the extremely difficult realization that something other than oneself is real."[21] Sueño entering English might help English speakers' dreams become less stereotypical and more traumatypical, making ecotopian futures more imaginable.

When celebrating his 2017 reelection as the mayor of Los Angeles, Eric Garcetti evoked his vision of the city as inclusive to all people, regardless of religion, legal status, ethnicity, or sexual orientation. Garcetti described this civic ideal as the "Los Angeles Dream."[22] Delinking Los Angeles from some national political imaginary newly entertaining toxic ideologies of exclusion, the mayor's phrase "the Los Angeles Dream" shifts toward the traumatypical, but the phrase "el sueño de Los Ángeles" would do so even more effectively. Given the meanings of the word sueño that this essay underlines, the phrase "el sueño de Los Ángeles" might suggest the city as a postcarbon metropolis open to dismantling sovereignty's traditional anthropocentrism in pursuit of multispecies coexistence.

As the American Dream becomes more stereotypical in Stiegler's sense of the word, el sueño americano may enact the traumatypical. In specific local or national forums, to insist on this phrase would enact resistance to the scapegoating of immigrants and challenge the vision of a United States walled off from Mexico. This vision already dreams the near future as ecodystopian, given the likelihood that climate change will in the coming decades force millions of Central Americans and Mexicans northward. The development of humane, environmentally sound policies and practices regarding these climate refugees will depend on the delimitation of sovereignty (Calderón), the free pursuit of knowledge (Sor Juana), and the transformative hope (Paz) that the word sueño invites U.S. speakers of English to embrace.

Admittedly, the ecotopian force of the word sueño suggested here involves a selective assemblage of meanings, and any single word's force is limited. Yet denialist messaging relies on the force of words, images, and dreams, as does the struggle in the Anthropocene for an ecotopian sueño of coexistence. For more and more humans and nonhumans, a future dreamt as fossil fueled—that is, as an extension of the present—may only arrive as a nonfuture: extinction. The fossil-fueled American Dream, a precipitate of the Great Acceleration, dreams the future as an extension of the most unsustainable aspects of the present. *El sueño americano, el sueño de Los Ángeles*, or a more local or individual sueño might dream the future as ecotopian.

However impatiently, however immodestly, however improbably, *tengo un sueño*. This sueño is impatient as it demands hospitality for and solidarity with the stranger. This sueño is immodest as it implies a contestation of the entire panoply of stereotypes operative in the fossil-fueled American Dream. This sueño is improbable as it envisions a future beyond what can seem all too probable: the Anthropocene becoming a fossil-fueled dead end.

As the American Dream, made "great" again, takes a dystopian turn that features the empowerment of climate change denial in the U.S. federal government and a supremacist and ecodystopian fantasy of walling the United States off from Mexico, welcoming sueño into English, specifically American English, would be a timely act. Rather than walling Spanish off or walling the word sueño in, English would host sueño as an inassimilable guest. Sueño would open interstices in English, allowing what might be to interlace with what is, giving ecotopian visions a chance to haunt English speakers' Anthropocene existence.

Nahual
ANOTHER PATH

NOTES

Dedication: For Melissa Santos.

1. See Will Steffen, Wendy Broadgate, Lisa Deutsch, Owen Gaffney, and Cornelia Ludwig, "The Trajectory of the Anthropocene: The Great Acceleration," *Anthropocene Review* 2, no. 1 (2015): 81–98.
2. Steffen et al., "Trajectory of the Anthropocene," 94.
3. Steffen et al., "Trajectory of the Anthropocene," 94.
4. James Truslow Adams, *The Epic of America* (Boston: Little, Brown, 1931), viii.
5. J. R. McNeill and Peter Engelke, *The Great Acceleration: An Environmental History of the Anthropocene since 1945* (Cambridge, Mass.: Harvard University Press), 128.

6. Matthew Schneider-Mayerson, *Peak Oil: Apocalyptic Environmentalism and Libertarian Political Culture* (Chicago: University of Chicago Press, 2015), 60.

7. Schneider-Mayerson, *Peak Oil*, 60.

8. John Archer, "The Resilience of Myth: The Politics of the American Dream," *Traditional Dwellings and Settlements Review* 25, no. 2 (2014): 11.

9. "North America: United States," World Factbook, Central Intelligence Agency, May 1, 2018, https://www.cia.gov/.

10. McNeill and Engelke, *Great Acceleration*, 5.

11. See Bernard Stiegler, "The Organology of Dreams and Arche-Cinema," in *The Neganthropocene*, trans. and ed. Daniel Ross (London: Open Humanities Press, 2018), 154–71.

12. Schneider-Mayerson, *Peak Oil*, 63.

13. *Bullitt*, dir. Peter Yates (1968; Warner Home Video, 1997), DVD.

14. Tom Cohen, "*Arche*-Cinema and the Politics of Extinction," *Boundary 2* 44, no. 1 (2017): 246.

15. Pedro Calderón de la Barca, *La vida es sueño/Life Is a Dream*, trans. and ed. Stanley Appelbaum (Mineola, N.Y.: Dover, 2002), 184.

16. Sor Juana Inés de la Cruz, "El sueño/First I Dream," in *Poems, Protest, and a Dream: Selected Writings*, trans. Margaret Sayers Peden (New York: Penguin, 1997), 78–129.

17. Gabriel Jackson, *The Spanish Republic and the Civil War, 1931–1939* (Princeton, N.J.: Princeton University Press, 1965), 300.

18. Octavio Paz, *El laberinto de la soledad y otras obras* (New York: Penguin, 1994), 49.

19. Paz, *Laberinto*, 48.

20. Paz, *Laberinto*, 48.

21. Iris Murdoch, "The Sublime and the Good," *Chicago Review* 13, no. 3 (1959): 51, quoted in Dale Jamieson and Bonnie Nadzam, *Love in the Anthropocene* (New York: OR, 2015), 204.

22. *PBS NewsHour*, March 8, 2017, PBS, https://www.pbs.org/.

Terragouge

Chris Pak

Pronunciation: te-ra-gowj

(tɛɹəgaʊdʒ)

Part of Speech: Verb

Provenance: Environmental science fiction

Example: Who profits? Not us. We sicken and die as they terragouge the land for the benefit of others elsewhere.

Terragouge highlights the significant contribution made by resource extraction, industrial farming, and urbanism to the human-induced transformation of Earth's planetary environment. The Anthropocene—whether dated to the Neolithic origins of agriculture, the ongoing Holocene extinction (the sixth mass extinction event), James Watt's plans for the steam engine (1784), or the widespread adoption of the steam engine in the early nineteenth century—denotes a geologic period inaugurated by new energy infrastructures that have enabled humankind to enact geological-scale transformations to Earth's surface. The extraction of coal was fundamental to driving the expansion of this infrastructure, and thus the Anthropocene is founded on a systematic expansion of what we might call terragouging. This entry examines the significance of the term terragouge for current and future discussions of resource extraction, energy, agriculture, and space colonization, contextualized by the history of the concept in science fiction.

Terragouge allows environmentalists, scientists, ethicists, economists, writers, and artists to make logical, discursive, and practical distinctions at the levels of intention and consequence between different modes of planetary adaptation.[1] Distinctions such as whether to transform the landscape for the purpose of human habitation, which is the goal of terraforming, or for the purpose of maximizing resource extraction, as in the case of terragouging, have become critical. The ability of humankind to alter the environment through technological means and at ever-increasing scales should encourage thought and planning for the best way to harness these abilities to ensure the continued integrity of Earth's biosphere for the flourishing of its inhabitants. As humankind's technological capacity and supporting infrastructures continue to grow, the prospect of irrevocably impairing living conditions on Earth for humans and nonhumans alike encourages environmental, ethical, scientific, and creative thought about how the Earth has historically been altered, how it continues to be transformed, and what the consequences of such modification might be. Whereas terragouging can be related to the concept (and violence) of the Anthropocene, the terms *terraforming* and *geoengineering* have a far broader reference.

The history of evolution on Earth involves multiple events that can be considered examples of terraforming (and not terragouging). The great oxygen event, in which the early colonization of the planet by cyanobacteria and the subsequent transformation of Earth's atmosphere from a carbon dioxide–rich environment to one containing enough oxygen for other forms of life to evolve, is a prototypical example of naturogenic terraforming. Such instances of nonhuman organisms effecting dramatic changes to Earth undermines the notion that humankind is exceptional in its ability to alter planetary environments. Indeed, this long history of transformation informs James Lovelock's Gaia hypothesis (1975) and undergirds other contemporary theories of planetary evolution.[2] According to many thinkers, only terraforming can ensure our species' survival in the face of the contemporary climate crisis.

DEFINITIONS

The term terragouging was coined by literary scholar and ecocritic Patrick D. Murphy. Murphy defines the term as the adaptation of planetary environments "to facilitate extraction of raw materials for earthly consumption."[3] Terragouge is a portmanteau derived by analogy with the term *terraforming*, which can be defined as the adaptation of planetary landscapes to facilitate their habitation by earthbound life. Terragouge combines the Latin noun *terra*, "earth," with the verb *gouge*, but it is as the gerund terragouging that the term most often appears. In this form, terragouging may refer to a broad range of activities that center on resource extraction.

The related term *terraforming* was invented by the science fiction writer Jack Williamson in his 1942 short story "Collision Orbit," although the concept is older. H. G. Wells, for example, portrays Martians who transform Earth's environments in an act of areoforming (the adaptation of other planets to facilitate habitation by Marsbound life) in his 1898 novel *The War of the Worlds*, and Han Ryner recounts in his 1909 short story, "A Biography of Victor Venturon" (first published in French and translated into English in 2011), the modifications that the eponymous scientist enacts on Earth's planetary environment in response to an impending global natural disaster.[4]

Terraforming and its counterpart on Earth, geoengineering, are subcategories of planetary engineering. Depending on its context, terragouging can likewise be considered a subcategory of planetary engineering at the same level as terraforming and geoengineering, or it can be considered a subset of terraforming.[5] The discourse of planetary adaptation—a discourse which once seemed futuristic, but is primed to move to the center of the climate change discussion—does not restrict these terms to humankind's contemporary abilities or contexts. Future endeavors, such as the mining of asteroids for raw materials—which is already being investigated by private companies such as Planetary Resources—would also constitute a form of terragouging. Likewise, should humankind establish interplanetary colonies or space stations—as Mars One and Elon Musk's Space-X plan to do—the definition of terragouging should be extended to include the extraction of raw materials for human consumption from outer space, whether or not that consumption only benefits communities on Earth. By underscoring the underlying motive for terragouging as the satisfaction of consumption on Earth, Murphy's definition draws attention to how humankind organizes and transforms matter around its needs and desires. Earth and human wishes are the touchstones around which the transformation of other planets is oriented.

Because the terms *terraforming* and *geoengineering* include a wide range of concepts and encompass many different types of planetary adaptation, the term terragouging allows further distinctions to be drawn between approaches to transforming nature. The utility of terragouging lies in its capacity to highlight how energy systems and the management of other natural resources are generally the primary motivation for transforming landscapes. Terragouging thus highlights the difference between the adaptation of planetary environments for the purpose of facilitating habitation by humankind and other terrestrial organisms, and the extraction of raw materials for consumption elsewhere. Naomi Klein identifies the operation of the Canadian tar sands as an example of terradeforming—that is, terragouging—because the landscape is not simply deformed but materially diminished.[6] Any act of terraforming can be construed as a planetary deformation, which harbors the implicit negative value judgment implied by the term *deforming*. Terragouge, by contrast, emphasizes the distinction between resource extraction that will exacerbate climate change and other modes of planetary adaptation, such as a welcome global project of deurbanization. For example, Laurence Manning, in his 1933 science fiction serialization "The Man Who Awoke," imagines a world that has deurbanized by depopulating urban centers and transforming the Earth into a pastoral landscape sparsely inhabited by a network of arboreal communities.[7]

Terragouging encompasses activity such as resource extraction in support of humankind's energy infrastructures, urbanism, and agro-industry. Given that carbon-based energy regimes are a significant cause of climate change, terragouging directly opposes terraforming insofar as the historical proliferation of wood, coal, oil, and gas as energy sources undermine Earth's capacity to support habitation by multiple forms of life. Just as terragouging relies on political, social, and economic decisions that have the capacity to reshape each of those domains, historical energy infrastructures come about through choices that reshape labor and the land in exploitative ways. The United Kingdom's Lancashire cotton industry is one example of such reconfigurations. It involved the application of steam-driven mechanized production for the spinning of wool and imports of cotton from plantations in North America. The transformation of the environment through agroindustry and factory farms is also a form of terragouging. These systems require resources such as nutrients, trace minerals and metals, water for food production, the manufacture of chemicals used in agricultural and factory farm processes, and resources for the genetic manipulation of plants and animals. Different modes of adaptation speak to different ways of valuing environments and their inhabitants.

In the case of terragouging, other organisms as well as nature at large are seen in instrumental terms as a means to an end: the maintenance of an economic system rooted in expanding consumption, primarily for a privileged subset of the human population. Historically, this mass consumption has not been restricted to capitalist economies but has also been a feature of other economic and political systems. Yet terraforming and terragouging are variations on one set of practices and are not radically separated or exclusive concepts. The character of any terraforming or terragouging program involves technical, social, and economic decisions that may be consistent with one another or significantly diverge in one or many respects. Indeed, terraforming and terragouging can in certain contexts be seen as identical if the transformation of another planet for the purpose of habitation is solely based on the extraction of raw materials for consumption by the agents of terraforming. Nevertheless, terraforming and terragouging often work at cross-purposes, as the narrator of Kim Stanley Robinson's novel *2312* (2012) concedes: "It was also true that metals and useful chemicals in lunar rock could be mined only by a deep strip-mining and processing of much of the lunar surface, which also made terraforming difficult. So large domed craters and tented areas alternated with cosmologically large mining pits, and each nation with a substantial lunar presence had an influx of raw materials."[8] Here terragouging as a form of "deep strip-mining" directly impedes efforts to transform Earth's moon into a habitable environment.[9] Nations that have staked a legal claim to portions of the lunar land own the extracted resources, which they transport to Earth to satisfy terrestrial demands for consumption. The overriding purpose of the terragouging of the moon is the maintenance of a capitalist economy on Earth, which directly impedes efforts to establish a viable ecologic and economic system on the moon that would enable its future habitation by currently earthbound life.

ENVIRONMENTALISM AND POPULAR CULTURE

Contemporary environmental struggles over resource extraction can be framed as a struggle against the terragouging of Earth. *Gouge* refers to the notches or grooves made in wood, but it also refers to an extreme act of violence that deprives one of sight. Colloquially, *gouge* refers both to the act of overcharging and to the mining of ore. The incorporation of the negatively shaded *gouge* and the allusive onomatopoeic quality of this verb, suggestive of circularity and brutality, offer a memorable and expressive term to organize resistance against a range of interventions into various landscapes.

The emergence of popular forms of resistance in opposition to new extractive technologies—such as hydraulic fracking, the 2015 kayak-tivist protest against Arctic drilling by Shell, the NoDAPL campaign against the Dakota Access Pipeline led by Native American water protectors in 2016 and 2017, and protests regarding the Canadian tar sands in Alberta—is an example of resistance to terragouging driven by alliances between multiple and heterogeneous groups. The term terragouging thus has the capacity to underscore the shared scientific, social, and economic bases of specific extractive events. However, it also has the capacity to connect resistance to extraction to resistance to other industrial processes that enact terragouging, such as agriculture, factory farms, fisheries, and mining. Agroindustrial interventions such as factory farming and the conjoining of biotechnology and agricultural monocultures pioneered by companies such as Monsanto and Syngenta illustrate how the extraction of raw organic materials from the Earth for consumption is conducted in terms of terragouging. For example, Indian environmentalist Vandana Shiva exposes how the Green Revolution's attempt to enhance agricultural yields in Asia also brought with it long-term environmental, economic, and public health problems for the populations inhabiting the areas directly affected by this shift in agricultural production processes.[10]

The utility of terragouging will increase in relevance as geoengineering for climate mitigation and planning for the human colonization of the solar system gain traction. A term such as terragouging would allow ethicists, economists, environmental scholars, and potential colonists to factor the ethics of resource extraction into decisions to transform cosmological nature, which also bear implications for the management of resources on Earth. Public–private partnerships are an increasingly popular model for imagining the colonization of the solar system, and plans for the mining of planetary bodies and asteroids grow apace. Companies such as Moon Express, Planetary Resources, and Bradford Space are built on a terragouging ethic, while business models based on astrotourism by companies such as Space-X, Virgin

Galactic, and Bigelow Aerospace are dependent on resource extraction from the solar system.[11] Mars One plans to establish a self-sufficient colony on Mars—a goal that could conceivably be achieved with the aid of the terragouging of Mars and a supporting infrastructure for the mining and transportation of resources from nearby asteroids. Given the tragic record of the exploitation of natural resources on Earth, it is imperative to establish an ethics that would inform and constrain further efforts to extract wealth from the solar system.

Interrogations of terragouging have been an important element of science fiction since the 1930s, and the theme has gained increasing popularity in wider culture since 2000 as a way to think about contemporary environmental issues. James Cameron's 2009 film *Avatar* is based on the conflict between colonizing humans who wish to terragouge the alien moon Pandora and the indigenous Na'vi, whose culture and material existence are threatened by these plans to turn their home into a mine.[12] Similar narratives appear in science fiction literature: Ursula K. Le Guin's 1972 novel *The Word for World Is Forest* likewise examines the ethics and politics of colonization and terragouging by portraying an indigenous people whose culture is intimately connected with a verdant forest landscape.[13] Duncan Jones's 2009 film *Moon* portrays the terragouging of Earth's moon for helium-3, a resource that allows Earth to meet its large demand for energy.[14] Ian MacDonald, in his 2015 novel *Luna: New Moon*, portrays the corporate terragouging of Earth's moon for helium-3 and other resources.[15] MacDonald shows how terragouging is made possible through the exploitation of all but the most powerful of the moon's population by corporations that exercise political and economic control on the moon.

Kim Stanley Robinson consistently explores issues of terragouging in his novels. His landmark Mars trilogy, comprising *Red Mars* (1992), *Green Mars* (1993), and *Blue Mars* (1996), portrays the colonization of Mars and the struggle between terragouging and terraforming models of planetary adaptation.[16] Robinson extends his reflection on the variety of environmental and political issues implicated in terragouging in his collection of short stories, *The Martians* (1999), as well as in more recent novels such as *2312* and *Aurora* (2015).[17] These novels offer complex visions of the competing ethical, political, and economic discourses that are a fundamental part of the decision to transform the landscape for anthropocentric ends and show how terragouging is a fundamental concept for thinking about anthropogenic climate change on Earth.[18]

Terragouge—an invaluable concept for discussing and responding to anthropogenic climate change—incorporates ethical, political, economic, scientific, social, and cultural concerns that are fundamental to contemporary human and nonhuman ecologies. Although the term is new, the concept has a long pedigree, and it has been developing within scientific and popular discourses for over a century. By offering a greater degree of discrimination between forms of adapting the planetary biosphere, terragouging can help us to explore modes of habitation and adaptation that would allow us to appropriately respond to the urgency of climate change. Robert A. Heinlein underscores this urgency in his 1966 novel *The Moon Is a Harsh Mistress* when a character challenges the terragouging of the moon by linking extraction to future suffering: "Every load you ship to Terra condemns your grandchildren to slow death."[19]

NOTES

1. Each of these groups has used terraforming in different ways to think about issues related to the environment. For example, ethicist and political philosopher Robert Sparrow has used terraforming as a thought experiment to explore an agent-based virtue ethics that is applicable to questions of environmental modification on Earth. Likewise, scientists and engineers such as James Oberg have considered environmental themes and issues as part of their theorization of terraforming. See Robert Sparrow, "The Ethics of Terraforming," *Environmental Ethics* 21, no. 3 (1999): 227–45.

2. See James Lovelock and Sidney Epton, "The Quest for Gaia," *New Scientist* 65, no. 935 (1975): 304–6. Additionally, for the role of plant–fungi symbioses in planetary

Total Liberation
ANOTHER PATH

adaptation, see Benjamin J. W. Mills, Sarah A. Batterman, and Katie J. Field, "Nutrient Acquisition by Symbiotic Fungi Governs Palaeozoic Climate Transition," *Philosophical Transactions of the Royal Society B* 373, no. 1739 (2018): 20160503; for algae's role in supporting the development of animal life, see Jochen J. Brocks, Amber J. M. Jarrett, Eva Sirantoine, Christian Hallmann, Yosuke Hoshino, and Tharika Liyanage, "The Rise of Algae in Cryogenian Oceans and the Emergence of Animals," *Nature* 548 (2017): 578–81; for the role of whales in reorganizing the structure and function of the world's oceans, see Joe Roman, James A. Estes, and Lyne Morissette, et al., "Whales as Marine Ecosystem Engineers," *Frontiers in Ecology and the Environment* 12, no. 7 (2014): 377–85.

3. Patrick D. Murphy, "The Non-alibi of Alien Scapes: SF and Ecocriticism," in *Beyond Nature Writing: Expanding the Boundaries of Ecocriticism*, ed. Karla Armbruster and Kathleen R. Wallace (Charlottesville: University of Virginia Press, 2001), 270.

4. See Jack Williamson, "Collision Orbit" (1942), in *Seventy-Five: The Diamond Anniversary of a Science Fiction Pioneer* (Michigan: Haffner Press Oak, 2004), 216–77; H. G. Wells, *The War of the Worlds* (1898), Project Gutenberg, http://www.gutenberg.org/; and Han Ryner, "A Biography of Victor Venturon," in *The Superhumans and Other Stories*, trans. Brian Stableford (Tarzana, Calif.: Black Coat Press, 2011).

5. Terraforming and geoengineering are becoming increasingly popular terms in climate science, news, politics, and culture. The Intergovernmental Panel on Climate Change 2014 report mentions geoengineering as a potential response to climate change, while writers such as Naomi Klein, George Monbiot, and Oliver Morton all discuss terraforming and geoengineering. Geoengineering has also provided the basis for movies such as *Snowpiercer* (2013) and *Man of Steel* (2013). See Naomi Klein, *This Changes Everything: Capitalism vs. Climate Change* (London: Allen Lane, 2014); George Monbiot, *Feral: Rewilding the Land, the Sea, and Human Life* (Chicago: University of Chicago Press, 2014); and Oliver Morton, *The Planet Remade: How Geoengineering Could Change the World* (Princeton, N.J.: Princeton University Press, 2015).

6. Klein, *This Changes Everything*, 139.

7. Laurence Manning, *The Man Who Awoke* (1933; reprint, New York: Ballantine, 1979).

8. Kim Stanley Robinson, *2312* (London: Orbit, 2012), 370.

9. In this example, the way they strip-mine the moon would impede terraforming. It is conceivable that strip-mining, given another planet with different characteristics (or even different approaches to strip-mining), might also help the terraforming of a planet based on its planetary features. Different science fiction texts place terraforming and terragouging in different relationships, and some texts might contradict the assertions of others. The example of the Alberta tar sands

is a good parallel. An example of the converse might be the moholes of the Mars trilogy, which use mining sites to create geological features that induce variations of heat in the atmosphere, which eventually aid in warming the planet (although this intervention is not unopposed in the narrative). Perhaps the clearest alternative would be the terragouging of an asteroid for water, which would be used in a terraforming project. Although the asteroid is terragouged into nonexistence, it provides a basis for the terraforming of planetary bodies in many narratives of terraforming.

10. Vandana Shiva, *The Violence of the Green Revolution* (London: Zed Books, 1993).

11. See Moon Express (http://www.moonexpress.com/), Planetary Resources (http://www.planetaryresources.com), Bradford Space (http://deepspaceindustries.com/), Space-X (https://www.spacex.com/), Virgin Galactic (https://www.virgingalactic.com/), and Bigelow Aerospace (http://www.bigelowaerospace.com/).

12. See *Avatar*, dir. James Cameron (20th Century Fox, 2009).

13. See Ursula K. Le Guin, *The Word for World Is Forest* (1972; New York: Berkley, 1976).

14. See *Moon*, dir. Duncan Jones (Stage 6 Films, Liberty Films, Xingu Films, and Limelight, 2009).

15. See Ian McDonald, *Luna: New Moon* (London: Gollancz, 2015).

16. See Kim Stanley Robinson, *Red Mars* (1992; reprint, London: Voyager, 1996), *Green Mars* (1993; reprint, London: Voyager, 1996), and *Blue Mars* (London: Voyager, 1996).

17. See also Kim Stanley Robinson, *The Martians* (London: Voyager, 2000) and *Aurora* (London: Orbit, 2015).

18. Space constraints prevent a comprehensive overview of the large number of terragouging narratives that abound. For further information on such narratives, see Ursula K. Heise, "Martian Ecologies and the Future of Nature," *Twentieth Century Literature* 57 (2011): 447–62; and Heise, "Reduced Ecologies: Science Fiction and the Meanings of Biological Scarcity," *European Journal of English Studies* 16 (2012): 99–112; Sylvia Kelso, "Tales of Earth: Terraforming in Recent Women's SF," *Foundation* 78 (2000): 34–43; and Chris Pak, *Terraforming: Ecopolitical Transformations and Environmentalism in Science Fiction* (Liverpool: Liverpool University Press, 2016).

19. Robert A. Heinlein, *The Moon Is a Harsh Mistress* (1966; reprint, London: Gollancz, 2001), 17.

Pronunciation: toe-tal li-ber-ay-shun

(tǝʊt(ǝ)l lɪbǝːreɪʃ(ǝ)n)

Part of Speech: Adjective and noun

Provenance: Animal rights and radical earth activists

Example: Only after we eliminate all forms of hierarchy, including racism and speciesism, will we achieve total liberation.

Total liberation is a phrase presently used by a small group of activists and scholars who identify with radical ecology and animal liberation movements. Though currently not in common use in English, it reflects and captures ideas, aspirations, and actions directed at confronting anthropogenic socioecological crises that have resulted in extraordinary harms across human and more-than-human communities around the globe. But total liberation is more than just a couple of words. It is also a framework for understanding and changing the world. It seeks to describe the world we live in while also offering a vision for transforming it. Consider the Boston Animal Defense League. As one activist publication reporting on the group states,

> The Boston Animal Defense League subscribes to the idea
> of total liberation and collective organization. All forms
> of oppression must be uprooted, from the exploitation of
> the Earth to the destruction of human and non-human

animals. We have to get to the root of our exploitation if we are to combat it effectively. For this reason, we constantly traverse movements that are often seen as separate struggles, including ecofeminism, deep ecology and workers' rights, and reject sexism, racism, homophobia and capitalism in the spirit of mutual aid. We are all in this movement together, in One Struggle, One Fight![1]

The total liberation frame comprises four components: (1) an ethic of justice and antioppression that is inclusive of humans, nonhuman animals, and ecosystems; (2) anticapitalism; (3) anarchism; and (4) an embrace of direct action tactics. These four components combine to produce a critique of the logic of domination—regardless of its source—and present a road map aimed at producing transformative and radical social change. In other words, in this time of massive social and ecological devastation, any efforts designed to make social change that leave existing social systems intact are counterproductive and unacceptable.

I stumbled on the notion of total liberation while doing research on radical ecology and animal liberation movements.[2] In addition to being focused on defending vulnerable nonhumans and ecosystems, it became clear that many activists in these social movements were also raising questions about whether their movements should also be struggles over social inequality more broadly. In other words, segments of these movements that had begun exclusively as radical environmental and animal liberation movements now expanded the terrain of struggle to include a more comprehensive view of social justice. Not only were they focused on defending nonhuman natures (i.e., animals and ecosystems), but they were also confronting oppression within human communities (i.e., racism, patriarchy, heterosexism, capitalism, state power, and empire). I wanted to know what sparked the development of this kind of intersectional movement. I found three motivations: (1) an increase in reports of threats to planetary sustainability and continued massive exploitation of nonhuman species and ecosystems associated with capitalist and state-based infrastructures; (2) frustration with the elitism, racism, patriarchy, and tactical reformism of mainstream animal rights and environmental movements; and (3) influences from other radical social justice movements, particularly movements of generations past that emerged from struggles among communities of the white working class and people of color. In order to deepen my understanding of what motivates total liberation activists' politics and practices, I explore each of these driving forces below.

The first influence on total liberation struggles is the ongoing series of threats to nonhuman species and ecosystems in the current epoch known as the Anthropocene. There is a consensus among leading scientists that damage to ecosystems over the last half century has been more severe than during any other time in human history. The health of coral reefs, fisheries, oceans, forests, and river systems declined dramatically, while climate disruption indicators, rates of species extinction, and air and water pollution rose significantly. At the same time, there has been an enormous increase in factory farming and industrial animal production, consumption, and experimentation that results in the slaughter of billions of nonhuman animals each year. This activity has created and accelerated negative impacts on ecosystems and nonhuman species. In responding to the varied local and global threats to ecosystems and nonhuman animal populations, many earth and animal liberation activists "hear" a "call" or "interpellation" from nonhuman natures that pushes them to defend ecosystems and nonhuman animals.[1] One activist involved in fighting a state highway construction project that threatened a water source and urban forest deemed sacred by indigenous peoples describes this dynamic: "If you open yourself to the land, the land will engage you. . . . We wanted to get direction from the land [as a] sacred space. . . . [so] people [would] go out into the land and to sit quietly with the land however long they wanted to, and listen and [be] open to what the land had to say to them about what we as humans could do to work with the land, to protect it."[4]

The second influence on total liberation struggles stems from the frustration among radical activists with mainstream ecological and animal welfare movements' orientation, values, and tactics. These generally include a lack of awareness of and commitment to antioppression politics, an embrace of state-centric and market-oriented "solutions" to socioecological threats, and a rejection of aggressive direct-action tactics in favor of peaceful, primarily lawful activism that seeks to educate and raise awareness rather than directly confront objectionable practices. Such mainstream activism includes direct cooperation with institutions known for threatening the health and well-being of ecosystems and nonhuman animal species. For example, the environmental NGO Conservation International partnered with the Newmont Mining corporation on Walmart's Jewelry Sustainability Value Network, which allegedly ensures that the gold sold in its stores is produced in socially and environmentally responsible ways.[5] Newmont is one of the world's largest gold mining companies; many of its facilities around the world have spawned outrage as a result of the ecological and public health concerns associated with leaking cyanide, mercury, cadmium, and arsenic in its mine tailings.[6] Radical activists see efforts like Conservation International's partnership as reformist or insider strategies because they reinforce existing structures of power that gave rise to anthropogenic socioecological crises in the first place. Total liberationists reject these approaches by challenging the elitism of ecological movements and supporting a greater willingness to risk personal freedom in pursuit of social change against dominant institutions and cultural practices.

The third influence on total liberation struggles comes from radical movements of the past that sought to break with the mainstream in order to pursue more direct and transformative changes in society. In their interviews and conversations with me, total liberationists invoked the Diggers and Levellers of Britain, slavery abolitionists, the Luddites, anarchist movements, the Industrial Workers of the World, movements for civil rights, Black Power, Puerto Rican independence, the American Indian Movement (AIM), the Weather Underground, Women's Rights and Gay and Lesbian Rights movements, ACT UP (AIDS Coalition to Unleash Power), and many others as their inspirations. These historical movements also endured state repression and saw many of their members jailed and imprisoned. Total liberationists take inspiration from these struggles when their own members face similar consequences. Today's radical ecology movements pursuing total liberation may be unique with respect to the particular emphases and combination of concerns they articulate, but they have much in common with movements of the past. They have all challenged dominant ideas, institutions, and modes of power, and they have paid dearly for it through state repression. All of these movements were vilified in their time by moderates and liberals yet were lionized decades or centuries later.

The emergence of the Earth and Animal Liberation Fronts (ELF and ALF) in the 1980s and 1990s marked a new stage in the evolution of ecological politics in the United States, Europe, Australia, Latin America, and Canada. This was a moment marked by a discourse of radical analysis and action rarely seen in environmental and animal rights movements until that point. By the late 1990s and early 2000s, segments of these movements were converging around new ideas, producing a broader discourse that linked ecology, social justice, antioppression, and animal liberation. They combined that discourse with direct actions like arson, sabotage/ecotage, animal rescue/liberation, and property destruction directed at a range of targets, including animal research laboratories, slaughterhouses, power lines, elite housing developments, ski resorts, fur farms, and industrial agricultural and poultry facilities. Through these efforts, total liberationists made visible their objections to the violence of capitalism, state power, multiple forms of oppression within human communities, speciesism (the belief that one species—in this case humans—is superior to another), and ecological destruction. Activists practicing direct action made conceptual links between harms to ecosystems and animals, and injustices facing other human beings. The following communiqué reports a direct action involving property destruction that reflects this link:

> Members of the Earth Liberation Front descended upon the
> Old Navy Outlet Center in downtown Huntington, Long
> Island [and] smashed . . . plate glass windows and one neon
> sign. This action served as a protest to Old Navy's owners'
> involvement in the clear cutting of old-growth forest in the
> Pacific Northwest. . . . Old Navy, Gap, Banana Republic care
> not for the species that call these forests home, care not for
> the animals that comprise their leather products, and care not
> for their garment workers underpaid, exploited and enslaved
> in overseas sweatshops.[7]

The idea of total liberation stems from a determination to understand and combat all forms of inequality and oppression. These movements organize and mobilize in favor of symbols, practices, and structures of equality and justice to do what social movements have always done: imagine and create a better world. This world would be premised on the idea that inequality and unfreedom in all their known manifestations should be eradicated, including inequalities imposed by humans on nonhuman species.

Total liberation can be said to extend concepts from social movement theory and women of color feminist theory into the nature–culture nexus—that space where humans and more-than-humans exist, collide, and sometimes collaborate.[8] Social movement theory tends to be human-centered and sees social change as occurring exclusively through the work of human actors in isolation from the nonhuman world.[9] Total liberation works to remedy this oversight by reflecting the political ecological reality that social change occurs not only when people interact with and among each other, but also with actors and elements from nonhuman domains (nonhuman animals, ecosystems, technology, paper, machines, tools).[10] Women of color feminist theorists coined the term *intersectionality* as a way of explaining the fact that gender, race, class, sexuality, and other forms of social difference work together in the experiences of individuals such that we cannot reduce oppression or privilege to a single social category. The total liberation frame goes a step further by suggesting that intersectionality is unnecessarily restrictive if it begins and ends with humans. Total liberationists contend that one cannot fully grasp the foundations and consequences of racism and heteropatriarchy (among humans) without also understanding speciesism and dominionism because the categories of nonhuman, human, race, gender, and sexuality are used to define one another.[11] The concept of total liberation reveals the complexity of various systems of hierarchy while also suggesting points of intervention, transformative change, solidarity, and coalition building within and across species boundaries.

The emergence of the total liberation framework is a response to socioecological inequalities. To realize its promise, earth and animal liberation movements must confront hierarchy both outside and within their ranks. The struggle is necessarily led by humans and is directed at other human beings, institutions, practices, and ideas such as social injustice, oppression, speciesism, dominionism, capitalism, the state, and imperialism. As biologist Barry Commoner notes, "The earth is polluted neither because man is some kind of especially dirty animal nor because there are too many of us. The fault lies with human society—with the ways in which society has elected to win, distribute, and use the wealth that has been extracted by human labor from the planet's resources. Once the social origins of the crisis become clear, we can begin to design appropriate social actions to resolve it."[12] Although total liberationists may be a statistical minority, many important aspects of the larger fight against oppressive social systems in society take place within these movements themselves as they grapple with the spectrum of various forms of inequality and violence that have traditionally permeated and haunted environmental and animal rights causes.

Many ideas at the core of total liberation are compatible with environmental justice (EJ) movements, and vice versa. In fact, the ideas that radical animal and earth liberation activists express in their public and internal movement conversations are almost entirely reflective of concepts contained at the heart of the principles of EJ—a sort of founding document of the U.S. EJ movement, penned in 1991. The EJ principles reflect opposition to racism, patriarchy, the excesses of the state and market forces, and ecological harm; they recognize the inherent worth of nonhuman natures and acknowledge the inseparability of humans and the more-than-human world, all of which are cornerstones of total liberation. The EJ and earth/animal liberation movements may oppose myriad forms of hierarchy, inequality, and institutional hegemonies for different historical reasons and with distinct emphases, but their focus on such practices suggests possibilities for cross-movement conversation, analysis, and collaborative action.

However, there are limitations within this emergent discourse. For example, in their work to confront the socially constructed hierarchies and divisions between humans and nonhumans, scholars and activists who embrace the concept of total liberation are invariably faced with the historically entrenched association of non-Europeans and women with a "state of nature," as highlighted in the writings of Enlightenment-era philosophers such as Jean-Jacques Rousseau, John Locke, and Samuel von Pufendorf. That is, every step toward fighting speciesism and anthropocentrism is also a step that comes dangerously close to what philosopher David Theo Goldberg calls "naturalism"—those theories of humanity that place non-Europeans and women in an inherent state of inferiority because of their supposed subhuman (read: "animal" or "natural") qualities.[13] Total liberationists run the risk of reintroducing or reproducing racism, heteropatriarchy, and anthropocentricism that arguably gave rise to contemporary socioecological crises. These activists seek to flatten the hierarchies and divisions between humans and nonhumans, which may sound like an admirable goal. For some groups, such as women and people of color, this is a problem because those hierarchies were already erased centuries ago by scholars, corporations, and state authorities who sought to support racial and gender inequalities that often linked those populations with devalued nonhuman animals. Arguably, we are left with that legacy today, and total liberationists often fail to appreciate that history. This is a major conundrum with no clear and easy resolution, but it has not deterred activists and scholars from enthusiastically exploring these ideas and politics.

Yet total liberation is a valuable loanword. It reveals the myriad creative, affective, material, discursive, generative, and sometimes destructive ways in which some of us work to address the everyday and existential threats associated with socioecological crises and the Anthropocene. Total liberation reflects the ways in which social change makers push beyond the conventions and boundaries of mainstream ecological politics to imagine and enact ecotopian dreams of freedom for all beings and things.

NOTES

1. "Trenches Spotlight: Boston Animal Defense League," *No Compromise* 29 (2006): 31.
2. See David N. Pellow, *Total Liberation: The Power and Promise of Animal Rights and the Radical Earth Movement* (Minneapolis: University of Minnesota Press, 2014).
3. See Paul Robbins, *Lawn People: How Grasses, Weeds, and Chemicals Make Us Who We Are* (Philadelphia: Temple University Press, 2007).
4. Anonymous activist interview with the author, October 2010.
5. Newmont Mining Corporation, "Newmont Selected to Partner with Wal-Mart and Conservation International in Responsible Mine-to-Market Jewelry Initiative," CSR Wire, July 15, 2008, http://www.csrwire.com/.
6. "Peru: Mine Protest Resumes," Associated Press, January 2, 2012.
7. Leslie James Pickering, ed., "March 5, 2001," in *The Earth Liberation Front, 1997–2002* (Portland, Ore.: Arissa Media Group, 2007), 28–29.
8. The concept of nature–culture allows us to think about the ways in which humans and the more-than-human world are inseparable, intertwined, and collectively

Apocalypso
ANOTHER PATH

collaborate to produce one another. In that way, we come to understand that the fate of each one is necessarily bound up with the other.

9. See Doug McAdam and Hilary Boudet, *Putting Social Movements in Their Place* (Cambridge: Cambridge University Press, 2012); and Francesca Polletta, *Freedom Is an Endless Meeting: Democracy in American Social Movements* (Chicago: University of Chicago Press, 2002).

10. Jane Bennett, *Vibrant Matter: A Political Ecology of Things* (Durham, N.C.: Duke University Press, 2009).

11. Speciesism is the idea that one species is superior to another. Dominionism is the specific form of speciesism that views humans as superior to all other animal species. This view of a human-dominant multispecies order is supported by a cultural framework that declares this arrangement to be morally just.

12. Barry Commoner, *The Closing Circle: Nature, Man, and Technology* (New York: Knopf, 1971), 176.

13. David Theo Goldberg, *The Racial State* (Malden, Mass.: Blackwell, 2002).

Pronunciation: wa-ter-shed di-sigh-pull-ship
(wɑtəɹʃɛd dɪˈsʌɪp(ə)lːʃɪp)

Part of Speech: Noun

Provenance: Ecotheology

Example: My church practices watershed discipleship
by actively caring for our region as an expression
of our faith.

*Do unto others downstream what you would
have those upstream do unto you.*

—WENDELL BERRY,
CITIZENSHIP PAPERS (2003)

A few years ago, some neighbors in the Big Elkin Creek watershed in
North Carolina got into a dispute about agricultural runoff in a shared
creek. The neighbors, a local farmer and another landowner, argued
about who was responsible for correcting downstream problems resulting from sediment erosion. Because both these landowners were Christians, an employee at the soil and water conservation district asked a
local Presbyterian church pastor, Reverend Stuart Taylor, if he would
mediate the dispute. Taylor agreed, and the parties eventually settled
on a resolution. As Taylor developed a relationship with the farmer
and offered some gentle education on riparian health, couched in the

context of caring for God's creation, the tobacco farmer did even more remediation work on his creek than Taylor and the neighbor requested. Using watershed discipleship as his approach, Taylor made an ally in creek restoration, as well as the broader work of caring for one's region as an expression of Christian faith. The farmer now encourages other neighbors to care for their portions of Big Elkin Creek. Taylor saw this reconciling work as an expression of his Christian faith, aimed at meeting the social and ecological needs of his community.

This story, situated in one local watershed, illustrates several aspects of the potential value that watershed discipleship can offer toward the construction of a future ecotopia, especially in these troubled times. First, Christian congregations form a built-in network of relationships, holding land and leadership positions in many communities, and so it is important to describe the work of environmental care within the framework of the Christian message. Second, pastors and church leaders are often respected members of a community, and many are trained in dispute resolution, which is useful when dealing with natural resource conflicts. Third, churches are already networked with other churches within their region and abroad and partner with one another to coordinate various projects and initiatives. In these ways, congregations, pastors, and faith-based networks provide a ready-made infrastructure equipped to negotiate disputes and engage in advocacy and activism. They can form partnerships to build climate change mitigation and adaptation strategies and enact watershed restoration projects at the local level. As niche-level change agents, they can help translate climate change warnings into action.[1]

I came to care about the environmental crisis through my Christian faith and awareness of the justice issues that humanity faces today. Christian groups and other faith communities have played a major role in past struggles for civil rights and equality, and about a decade ago I began to realize that most of the social justice issues of our time relate to climate change, pollution, and natural resource conflicts. I am not the only person of faith recognizing the importance of environmental concerns; a major "greening of religion" has been occurring in faith communities in recent decades, with individuals, congregations, and denominations recognizing the problems of environmental degradation and climate change and recovering textual and historical resources for caring for the planet.[2] This is by no means indicative of a wholesale shift in the American Christian church, however. While at least 70 percent of Americans label themselves as Christian, according to a study by the Pew Research Center, those Americans who are most "alarmed" about climate change tend to be the least religious.[3] Increasingly higher percentages of religious people can be found in categories of those who are "disengaged," "doubtful," and "dismissive" of climate change.[4] Because of the number of Americans who are Christian and their doubts about anthropogenic climate change, it is critically important to communicate about contemporary environmental issues in their language.

Though the Christian tradition supports care for the Earth, many Christians are not acting in ways that will sustain comfortable or even livable conditions for many of the planet's existing species. Watershed discipleship provides a framework for environmental action based in Christian values. The kinds of actions that are already being inspired by watershed discipleship include adaptation and mitigation strategies to combat climate change; developing and supporting local food systems; and advocating for environmental justice. Individuals and congregations plant pollinator gardens filled with native species; partner with environmental groups; practice permaculture on church properties or unused urban spaces; coordinate community gardens or community-supported agriculture endeavors (CSAs); and engage in activism around farmworker justice, fossil fuel divestment, water rights, and other environmental justice issues. They model more sustainable lifestyles by retrofitting church buildings, installing solar panels, providing locations to reuse or recycle, hosting farmer's markets, and organizing communal living spaces with a focus on sustainability. In addition, watershed discipleship practitioners often incorporate themes of care for God's creation and watershed awareness into their worship services and offer educational opportunities such as workshops and green job training.

Coined by activist theologian Ched Myers around 2010, the concept of watershed discipleship integrates scientific knowledge about ecosystem services with stories and ideas already present in the Christian tradition.[5] Practitioners' acts of restoration and re-placement are tied to themes found in the Bible, giving weight and meaning to environmental actions for Christians. A watershed is the area where all the water falling in the region drains to a common outlet such as a stream, lake, or river; watersheds are divided by ridgelines, the other side of which flows to a different outlet. Watershed discipleship, therefore, envisions the watershed as a scalable unit of care: one's smallest local watershed may form around a stream in one's neighborhood, scaling up to a larger watershed based around a large river, and ultimately flowing out into the ocean, linking all watersheds. Thinking bioregionally encourages Christians to become environmentally conscious by starting from their own grounded location, while being aware of their impact on the entire biosphere. Although the Christian tradition contains many resources for environmental care, Christianity has been critiqued as a placeless religion of empire, encouraging the "watershed conquest" of colonialism, and has historically been complicit in Western civilization's degradation of natural resources.[6] Conquest of the land is a prevalent theme in parts of the Bible, such as the people of Israel taking over the Promised Land in the books of Joshua and Judges. Therefore, watershed discipleship attempts to acknowledge the ways Christianity has been used to legitimize conquest while emphasizing that this is not the only or the most prevalent story within the Judeo-Christian tradition. Care for the land is also a theme that weaves its way through the entire biblical narrative.

For the American church, and churches around the world that have bought into the colonialist version of Christianity, moving toward a sustainable future will require a great deal of work, starting with an acknowledgement of the reality of anthropogenic climate change and environmental degradation. It will be important for Christians to take responsibility for their complicity in creating these problems. To start, we can highlight portions in scripture that tell a different story.[7]

CHRISTIAN COMPLICITY IN THE
WATERSHED MOMENT OF ECOLOGICAL CRISIS

In 1967, historian Lynn White Jr. famously critiqued Christianity's role in propping up Western cultural and environmental imperialism, prompting theologians to respond by articulating what is now known as ecotheology: noting places in theology and Christian history that have contributed to an unhealthy anthropocentrism and recovering an ecological paradigm within the Christian scriptures. Moving from critique to action, however, continues to be a challenge. Watershed discipleship critiques interpretations of Christianity that prop up imperialism and conquest and translates theory to action through three foundational premises: (1) it considers the *watershed moment of ecological crisis* in which we find ourselves; (2) watershed discipleship encourages Christians to live as *disciples in our watersheds*, followers of Jesus Christ who live in and are connected to local humans and nonhumans; and (3) it encourages us to be *disciples of our watersheds*, listening to and learning from our bioregion to learn more about God through creation.

The concept of watershed discipleship contains an awareness of intersecting injustices around race, gender, and colonialism by drawing in the added dimension of the environment. Myers calls watershed discipleship a "critical ecotheology," and grounds its arguments for just action in the Christian faith.[8] In the tradition of the biblical prophets, watershed discipleship calls the faithful to a more complete expression of their faith, which includes right treatment of the land, animals, and marginalized people, rather than simply following church laws and rituals.[9] In this watershed moment of ecological crisis, watershed discipleship practitioners communicate the reality of anthropogenic climate change in Christian language, remind Christians of our shared story, and invite people to begin living in a way that is consistent with the biblical vision of love through justice. This holistic understanding of how God invites people to live can be summed up by the Hebrew term *shalom*, a multidimensional peace that includes reconciled relationships with God and other people as well as the land and nonhuman creatures.[10]

As a result of misinterpretations of Christian scripture, the creation story appears to give human beings dominion over the rest of creation.[11] This contributes to the perspective of some Christians that resources are here solely for our use. Moreover, the severing of the Earth's systems from Christian concerns is sometimes supported by Jesus's emphasis on his kingdom being not of this world and the universality of the claim of one God, accessible to all people, with instructions to Jesus's disciples to go into all the world to spread the good news.[12] Because of these and other scriptural texts, the Christian cosmology has often been envisaged as a hierarchy, with people (especially men) closest to God. These passages have also sustained a dualistic conception of reality, where the spiritual world is closer to godliness and the material world is evil.[13]

One of the most ecologically consequential policies espoused by the church was the Doctrine of Discovery, which was based on a series of papal bulls in the fifteenth century and used as substantiation for Native American land loss in legal cases as recently as 2005.[14] This doctrine legitimized colonial claims to an entire watershed under the principle of contiguity of that waterway. When a European explorer discovered the mouth of a river, the explorer could claim ownership of that entire watershed for his country—and the Roman Catholic Church—

if the current inhabitants were not Christians.[15] This led to the conquest and exploitation of entire watersheds and the people who lived in them. This conflation of the spread of the Christian message with Western imperialism legitimized injustice toward indigenous people, whose land was confiscated as a result of a misguided Eurocentrism that claimed "manifest destiny" for colonists to lay claim to the New World's land and resources, as well as injustice toward Africans, whom Europeans felt it was their duty to "civilize."

Though conquest-oriented interpretations of Christianity have been used to defend colonialism and justify the extraction and overuse of natural resources, the overarching themes present in scripture tell a different story. The books of the Jewish Law (Torah) focus on the covenant between God and Abraham regarding land. However, this promise of land is not an absolute right to the land and its resources without question. On the contrary, theologian Ellen F. Davis points out that the covenant is based on proper use (stewardship) because the land ultimately belongs to God.[16] Proper use of the land includes just treatment of the people who live there, as well as nonhuman animals and plants, creating a rhythm of work and rest that is healthy for people and the ecosystem.[17]

Watershed discipleship practitioners encourage "reinhabiting" the watershed by advocating that Christians become disciples in their watersheds as well as disciples of their watersheds. Practicing discipleship in our watersheds starts from the recognition that God chose to be incarnated into the body of Jesus in a specific place and time. This helps break down the idea that the material world is the opposite of or incompatible with the spiritual world: God became embodied, connected to an ecosystem, and was intimately involved with the justice issues of his day. While Christians are sometimes derided for being so focused on heaven that they are no earthly good, Jesus's life tells a story of good news for the poor and disenfranchised in this life. He invited his followers into a communal practice of faith in a God who cared about them as embodied beings, not simply their souls—a point that has been lost in some recent iterations of the Christian message.

Theologians sometimes refer to the "book of created nature" as another source of wisdom in addition to the Bible because it provides metaphors and learning opportunities through which people can learn about God.[18] Watershed discipleship advocates become disciples of their watersheds through learning about the social, economic, and environmental issues facing their region. This occurs as we allow other species and Earth systems to teach us, to be our rabbis, and combine this wisdom about ecological systems with what we also know about social and economic justice.

WATERSHED DISCIPLESHIP IN ACTION

The theme of water flows throughout scripture, particularly in the practice of baptism. The many conflicts regarding clean drinking water in recent years provide fertile ground for partnerships related to water justice. Seeing access to clean water as a basic human right, watershed discipleship helps provide a moral framework for advocacy and activism aimed at changing unjust laws. For example, Reverend Bill Wylie-Kellermann of Detroit, Michigan, practiced watershed discipleship when he engaged in civil disobedience in 2014, blocking trucks sent to shut off water to Detroit's residents. He and other members of the "Homrich 9" advocated for a water affordability plan based on income.[19]

Todd Wynward practices watershed discipleship by leading wilderness treks and learning to grow food sustainably in the arid climate of New Mexico. Wynward founded an organization called Taos Initiative for Life Together, and he combines an intimate knowledge of the land with spiritual depth based on stretches of time spent in the wilderness. He has focused on building relationships with and learning traditional skills from the Taos Pueblo people, following their lead, and standing with them in disputes regarding access to land, water, and sacred places. Wynward has made his home into a sustainable projects incubator where Christians can learn about watershed discipleship, spend time in community, and have time and space with minimal financial burdens so as to launch sustainability projects.[20]

Watershed discipleship is also enacted through practicing rituals and incorporating liturgical elements related to the community's watershed. Many who care about the environment find themselves overcome with guilt about their complicity in exacerbating environmental ills, and the weight of this knowledge can become paralyzing.[21] Christianity, with its rituals of lament, confession, and repentance, can offer space for individuals to grieve, confess their complicity, and make a commitment to changing their behavior. One way to begin connecting people with awareness of their watersheds is through Christian rituals. Congregations that use local waterways to perform baptisms, or local wine and bread for communion, naturally begin to attend to the cleanliness of the water and the way the ritual elements are produced.

People of faith participate in communities that form networks across a region, with denominational ties crossing the nation and globe. Actions taken within one specific watershed can be shared with those in other regions, sparking similar actions in other parts of the world. Shared faith combined with shared environmental concern can encourage partnerships for activism, the cultivation of more sustainable systems, and the implementation of restoration and mitigation projects. In my own region, near Portland, Oregon, Christian communities espousing watershed discipleship have partnered with 350PDX (a chapter of 350.org) to protest coal trains and terminals, advocate for the city of Portland to divest from fossil fuels, and provide gathering spaces where community groups can learn about and plan actions to address regional environmental concerns.

Many churches own land, and individuals and congregations practicing watershed discipleship are finding ways to creatively partner with others to utilize church land to contribute to developing food sovereignty and healthy, local food systems. Reverend Nurya Love Parish felt inspired to create a hub for the Christian Food Movement as an expression of watershed discipleship. In 2015, she developed a website that now lists hundreds of organizations and congregations working to provide access to healthy food in their communities. Parish started Plainsong Farm, an organic farm that provides education on sustainable farming and gardening, offers hospitality, sells CSA shares, and allows CSA members to purchase extra shares to give to those who can't afford them. The farm and the local Episcopal diocese are doing this work as a nonprofit ministry, caring for their community and land in tangible ways.[22]

Because Christians form a sizable percentage of the global population, and because Christianity in combination with Western empires has had a major impact on the way the land and its inhabitants have been treated in the last two millennia, expressing environmental care as an integral part of the Christian faith could help move our global practice toward a more sustainable future. Watershed discipleship helps articulate current environmental concerns and encourages action based on the Christian tradition and within the context of a unit of care (the watershed) that is within the scope of the human imagination and connects one's local actions to the entire world. Practicing watershed discipleship means spending time getting to know and love the watershed, connecting this knowledge to liturgical rituals, and partnering with others to address shared concerns. It also means communicating about one's work with those in other watersheds, in the knowledge that watershed discipleship will look different in each particular context, with the faith that Christian communities will be inspired by and learn from what is occurring in other regions.

NOTES

1. See Heila Lotz-Sisitka, Arjen E. J. Wals, David Kronlid, and Dylan McGarry, "Transformative, Transgressive Social Learning: Rethinking Higher Education Pedagogy in Times of Systemic Global Dysfunction," *Current Opinion in Environmental Sustainability* 16 (2015): 73–80; and Elise Amel, Christie Manning, Britain Scott, and Susan Koger, "Beyond the Roots of Human Inaction: Fostering Collective Effort toward Ecosystem Conservation," *Science* 356 (2017): 275–79.

2. Bron Raymond Taylor, "The Greening of Religion Hypothesis (Part One) from Lynn White, Jr. and Claims that Religions Can Promote Environmentally Destructive Attitudes and Behaviors to Assertions They Are Becoming Environmentally Friendly," *Journal for the Study of Religion, Nature and Culture* 10, no. 3 (2016): 268–305.

3. G. Smith, "Religious Landscape Study," Pew Research Center, Religion and Public Life, 2014, http://www.pewforum.org/.

4. Connie Roser-Renouf, Edward Maibach, Anthony Leiserowitz, Geoff Feinberg, and Seth Rosenthal, "Faith, Morality and the Environment: Portraits of Global Warming's Six Americas," Yale Program on Climate Change and Communication, January 19, 2016, http://climatecommunication.yale.edu/.

5. Ched Myers, ed., *Watershed Discipleship: Reinhabiting Bioregional Faith and Practice* (Eugene, Ore.: Cascade Books, 2016).

6. Katerina Friesen, "Watershed Discipleship as Home Mission: Toward a Constructive Paradigm of Repentance," *Missio Dei* 5, no. 2 (2014), http://missiodeijournal .com/; and Lynn White Jr., "The Historical Roots of Our Ecological Crisis," *Science* 155 (1967): 1203–7.

7. Ellen F. Davis, *Scripture, Culture, and Agriculture: An Agrarian Reading of the Bible* (Cambridge: Cambridge University Press, 2008).

8. Myers, *Watershed Discipleship*, 6.

9. For example, Hosea 4:1–3.

10. Randy Woodley, *Shalom and the Community of Creation: An Indigenous Vision* (Grand Rapids, Mich.: Eerdmans, 2012).

11. Genesis 1:26–29.

12. John 18:36 and Matthew 28:18–20.

13. Whitney Bauman, *Theology, Creation, and Environmental Ethics: From Creatio Ex Nihilo to Terra Nullius* (New York: Routledge, 2014); and Heather Eaton, *Introducing Ecofeminist Theologies* (London: Bloomsbury T&T Clark, 2005).

14. *City of Sherrill v. Oneida Indian Nation of N.Y.*, 544 U.S. 197 (U.S. Supreme Court 2005).

15. Friesen, "Watershed Discipleship."

16. Davis, *Scripture, Culture, and Agriculture*.

17. Exodus 23:10–12; see also Walter Brueggemann, *Sabbath as Resistance: Saying No to the Culture of Now* (Louisville, Ky.: Westminster John Knox Press, 2014).

18. Saint Augustine, Sermon 68, in *The Works of St. Augustine: A Translation for the 21st Century*, ed. John E. Rotelle, trans. Edmund Hill, 40 vols. (Brooklyn, N.Y.: New City Press, 1991); *Part 3—Sermons*, vol. 3: *Sermons 51–94*.

19. Holly Fournier, "Water Activists Speak Out Ahead of Court Case," *Detroit News*, November 18, 2015; and Candice Williams, "Trial Dismissed against 'Homrich 9' Water Protesters," *Detroit News*, June 21, 2017, https://www.detroitnews.com/.

20. Todd Wynward, *Rewilding the Way: Break Free to Follow an Untamed God* (Harrisonburg, Va.: Herald Press, 2015).

21. Chris Johnstone and Joanna Macy, *Active Hope: How to Face the Mess We're in without Going Crazy* (Novato, Calif.: New World Library, 2012).

22. Nurya Love Parish, "Yes, There Is a Christian Food Movement," Christian Food Movement, March 30, 2015, http://christianfoodmovement.org/.

Contributors

Sofia Ahlberg is associate professor in the English department at Uppsala University, Sweden, and a research fellow at La Trobe University, Australia. Her first monograph, *Atlantic Afterlives,* is about the impact of digital information on literature.

Randall Amster is codirector and teaching professor in environmental studies and the Program on Justice and Peace at Georgetown University. His recent books include *Peace Ecology* and *Anarchism Today* and the coedited volume *Exploring the Power of Nonviolence: Peace, Politics, and Practice.*

Brent Ryan Bellamy studies and teaches science fiction, American literature and cultures, and energy humanities. He has coedited *Materialism and the Critique of Energy* and a 2018 special issue of *Science Fiction Studies* on "Climate Crisis." He is writing a book on postapocalyptic novels during the age of U.S. decline.

Cherice Bock is a PhD candidate in environmental studies at Antioch University New England, where she edits the journal *Whole Terrain.*

Charis Boke completed her PhD in anthropology at Cornell University. Her research with herbalists and community organizers in New England is situated athwart science studies, medical anthropology, and environmental anthropology.

Natasha Bowdoin is assistant professor of painting and drawing at Rice University. She has been awarded residencies at the Core Program, Houston, Texas; the Roswell Artist-in-Residence program, Roswell, New Mexico; and the Bemis Center for Contemporary Art, Omaha, Nebraska. Her art has been exhibited at Mass MoCA (Massachusetts), the Moody Center for the Arts (Houston), the Savannah College of Art and Design Museum (Georgia), the CODA Museum (the Netherlands), the Southeastern Center for Contemporary Art (North Carolina), the Portland Museum of Art (Maine), the Cue Art Foundation (New York), and Extraspazio (Italy). She is represented by Monya Rowe Gallery (New York) and Talley Dunn Gallery (Dallas).

Kira Bre Clingen is a master of landscape architecture student at Harvard University's Graduate School of Design. She holds degrees in ecology and evolutionary biology, as well as in environmental policy studies and Asian studies. She was a 2016–17 Thomas J. Watson fellow.

Lori Damiano is assistant professor of animated arts at the Pacific Northwest College of Art. She has exhibited her paintings and screened her films nationally and internationally, including shows at New Image Art in Hollywood, Yerba Buena Center for the Arts in San Francisco, Fleisher-Ollman in Philadelphia, Mulherin + Pollard in New York City, and The Narrows in Melbourne, Australia. She has been commissioned to paint murals in Portland, Laguna Beach, Tokyo, Berkeley, and Guatemala. She served as chair of the animation department at the California State Summer School for the Arts for nine years and is on the Resource Council for the Independent Publishing Resource Center.

Nicolás de Jesús was born in an Aztec-Náhuatl indigenous community in Ameyaltepec, Mexico (Guerrero State). As a child, he learned from his parents how to use paintbrushes and colors to express our reality. Because he grew up unhappy with the reality of the discrimination and repression of his people, which has continued for more than five hundred years, art became a means for him to push out this unhappiness and to imbue with conscience other beings who do not dare cry out their pain. His art has been shown internationally in France, Canada, and Indonesia, as well as several cities in the United States. In Chicago, he collaborated with other artists to establish the Taller Mexicano de Grabado (Mexican Printmaking Workshop).

Jonathan Dyck is an illustrator and designer based in Winnipeg, Manitoba—Treaty 1 territory. His illustrations have been featured in a variety of publications, including *The Walrus*, *Reader's Digest*, *Maisonneuve*, and *Alberta Views*. He designed and illustrated the anthology *Unsettling the Word: Biblical Experiments in Decolonization*. His art can be viewed at jonathandyck.com or on Instagram (@jandrewdyck).

John Esposito is professor in the School of International Liberal Studies at Chukyo University in Nagoya, Japan, where he teaches mass media, global studies, and critical discourse analysis.

Rebecca Evans is assistant professor of English at Winston–Salem State University. Her writing has been published in *Resilience*, *Women's Studies Quarterly*, *Cambridge History of Science Fiction*, *Los Angeles Review of Books*, and *Public Books*.

Allison Ford is a doctoral candidate in sociology at the University of Oregon. She holds degrees in sociology and international environmental policy.

Carolyn Fornoff is assistant professor of Latin American literatures and cultures at the University of Illinois at Urbana–Champaign. Her recent writing has been published in *Revista de Estudios Hispánicos*, *Istmo*, *Mexican Literature in Theory*, and *The Cambridge History of Latin American Women's Literature*.

Michelle Kuen Suet Fung's oeuvre revolves around a grand narrative of a dystopian world set in the year 2084. Her works present a fictional geopolitical map of a bizarre future, one impacted by changes in the Anthropocene. She has exhibited internationally and has participated in artist residencies at Banff Centre, Canada; Island Institute, Alaska; and Art Omi, New York (as recipient of the Cecily Brown Fellowship). Her works have received awards including 50 Best Books for Secondary Students, Hong Kong Professional Teachers' Union (2018), Young Writers' Debut Competition, Hong Kong (2017), the Grotto Award, Hong Kong Baptist University (2015), and Award of Excellence, Fourth Greater China Illustration Awards (2012 and 2016).

Andrew Hageman is associate professor of English at Luther College in Decorah, Iowa. He has published articles on diverse topics, including David Lynch and Chinese cinema, and how to teach Björk's *Biophilia* alongside her correspondence with Timothy Morton. He coedited the 2016 issue of *Paradoxa* on "Global Weirding" and currently embraces his love of *Twin Peaks* as a staff writer for *Blue Rose Magazine* and *25YL: Damn Fine TV/Film*.

Michael Horka is a PhD candidate in American studies at George Washington University. His dissertation examines the relationship of desire to the scales of climate change found in late twentieth- and early twenty-first-century science, speculative, and climate fiction.

Yellena James is an artist/illustrator in Portland, Oregon. She is known for her colorful and instantly recognizable organic landscapes. Preferring pen and ink, gouache and acrylic, she combines complex abstract forms into dazzling images. She has participated in shows around the United States and overseas, including solo exhibitions at Stephanie Chefas Projects (Portland, Oregon), Giant Robot (San Francisco and Los Angeles), the Here Gallery (Bristol, U.K.), and the

Hijinks Gallery (San Francisco). She has done illustration work for the brands Anthropologie, La Mer, Crate & Barrel, Pyrex, and Relativity Media.

Andrew Alan Johnson is assistant professor of anthropology at Princeton University. His book *Ghosts of the New City* takes a historical look at the idea of the city in northern Thailand.

Jennifer Lee Johnson is assistant professor of anthropology at Purdue University. She has published in *Comparative Studies of South Asia, Africa, and the Middle East* and *Aquatic Ecosystem Health and Management*, as well as in the edited volumes *Handbook of Sustainability and Social Science Research*, *Subsistence under Capitalism: Historical and Contemporary Perspectives*, and *Landscape, Environment, and Technology in Colonial and Postcolonial Africa*.

Melody Jue is assistant professor of English at the University of California–Santa Barbara. Drawing on the experience of becoming a scuba diver, her monograph *Wild Blue Media: Thinking through Seawater* develops a theory of mediation specific to the ocean environment.

Jenny Kendler is an interdisciplinary artist, environmental activist, naturalist, and wild forager. She has exhibited at Storm King, MCA Chicago, Pulitzer Arts Foundation, Albright-Knox, MSU Broad Museum, California Academy of Sciences, the Chicago Biennial, and the Kochi-Muziris Biennale, and has been commissioned to create public projects for urban conservatories, remote deserts, and tropical forests She is co-chair of artist residency ACRE, is a member of artist collective Deep Time Chicago, and since 2014 has been the first artist in residence with environmental nonprofit NRDC. She and an interdisciplinary team were awarded an Andrew W. Mellon Foundation grant for her community co-created project *Garden for a Changing Climate*. More information is available at jennykendler.com.

Yifei Li is assistant professor of environmental studies at NYU Shanghai and global network assistant professor at NYU. His research examines environmental governance in China, focusing on state power, urban sustainability, and disaster resilience. He has received research support from the U.S. Department of Agriculture, the National Science Foundation, the University of Chicago Center in Beijing, and the China Times Cultural Foundation. His recent work has been published in *Current Sociology, Environmental Sociology*, and *Journal of Environmental Management*.

Nikki Lindt, a painter, has created and produced an environmental multimedia opera *The Noahs* for Amsterdam's Pompoen Theater. She received the Environmental Cultural Award from the Dutch Environmental Protection Agency in Amsterdam. After moving to Brooklyn, where she now resides, she co-curated two environmental-themed exhibitions titled "Natural Reaction." She has received the Pollock Krasner Grant, the Mama Cash Grant, and the Basis Grant from Fonds BKVB, Netherlands Foundation for Visual Art. She was nominated for the Anonymous Was a Woman Award. Her work has been covered by *Art Daily*, Huffington Post, the *Herald Tribune*, Asian Scientist, De Volkskrant, and NYMagazine. She is represented by Robischon Gallery and Heskin Contemporary.

Anthony Lioi is associate professor of liberal arts and English at the Juilliard School in New York. He is the author of *Nerd Ecology: Defending the Earth with Unpopular Culture*. He is now writing a book about metahumanism.

Maryanto received formal training in the visual arts at the Institute of the Arts, Yogyakarta. He has completed residencies at the Rijksakademie van beeldende kunsten, the Netherlands, and the Escuela de Orient programme at Casa Asia, Barcelona, Spain. He exhibited at the "Discoveries" section of Art Basel Hong Kong; Bazaar Art Jakarta; and Art Stage Singapore. He participated in the Setouchi Triennale;

the Yogyakarta Biennale, "Hacking Conflict: Indonesia Meets Nigeria"; and the Jakarta Biennale, "Neither Forward Nor Back." His recent projects include the ACC-Rijksakademie Dialogue and Exchange and "Europalia" at the Bozar Centre for Fine Arts in Brussels.

Janet Tamalik McGrath grew up between Inuit traditional and modern influences in the 1970s Canadian Arctic. A fluent speaker of Inuktut and lifelong language proponent, her doctoral work at Carleton University challenged Canadian academic norms by foregrounding Inuktut while also creating a unique bridge between epistemologies. Her research was published in *Qaggiq Model.*

Pierre-Héli Monot is a German Research Foundation postdoctoral fellow in comparative literature at the Ludwig Maximilian University of Munich. He received his PhD from the Humboldt University of Berlin and has held visiting research appointments at Brown University, Harvard University, the University of Oxford, and King's College London. His study of abolitionism and romantic hermeneutics, *Mensch als Methode: Allgemeine Hermeneutik und partielle Demokratie* (Man as a method: General hermeneutics and partial democracy), was published in 2016.

Susa Monteiro was born in 1979. She worked in theatre and cinema until 2009. Currently, she works exclusively in illustration and comics, and creates editorial illustrations for various Portuguese newspapers, books, and posters.

Daehyun Kim (Moonassi) was born and works in Seoul, Korea. His work explores the disagreement between minds and the endless inner conflict within us. More information may be found at moonassi.com.

Kari Marie Norgaard is associate professor of sociology and environmental studies at the University of Oregon. She has published on climate change, tribal environmental justice, gender, race, and risk perception. She is the author of *Living in Denial: Climate Change, Emotions and Everyday Life* and *Salmon Feeds Our People* (forthcoming).

Karen O'Brien is a professor of sociology and human geography at the University of Oslo, Norway. She leads the AdaptionsCONNECTS project, which looks at the relationship between climate change adaptation and transformations to sustainability.

Evelyn O'Malley is a lecturer in drama at the University of Exeter. Her monograph, *Weathering Shakespeare: Audiences and Outdoor Performance*, is forthcoming in Bloomsbury Academic's Environmental Cultures series.

Robert Savino Oventile is associate professor of English at Pasadena City College. He has published essays and book reviews in *Postmodern Culture, Jacket, symplokē,* and *Chicago Quarterly Review*. His poetry has appeared in *New Delta Review, Upstairs at Duroc,* and *Denver Quarterly*. He is the author of *Impossible Reading: Idolatry and Diversity in Literature* and *Satan's Secret Daughters: The Muse as Daemon*.

Chris Pak is the author of *Terraforming: Ecopolitical Transformations and Environmentalism in Science Fiction*. Since 2016 he has been subeditor of the Medical Humanities blog, and he was an editor of the Science Fiction Research Association's *SFRA Review*.

David N. Pellow is Dehlsen Chair and professor of environmental studies and director of the Global Environmental Justice Project at the University of California, Santa Barbara. His most recent books include *What Is Critical Environmental Justice?, Total Liberation: The Power and Promise of Animal Rights and the Radical Earth Movement* (Minnesota, 2014), and, with Lisa Sun-Hee Park, *The Slums of Aspen: Immigrants vs. the Environment in America's Eden*.

Andrew Pendakis is associate professor of theory and rhetoric at Brock University and research fellow at the Shanghai University of Finance and Economics. His essays on topics ranging from dialectics to plastics have been published in *South Atlantic Quarterly, Jacobin,*

Third Text, E-Flux, and *Critical Inquiry.* He is coeditor of *Contemporary Marxist Theory: A Reader* and the *Bloomsbury Companion to Marx.*

Kimberly Skye Richards is a PhD candidate in performance studies at the University of California–Berkeley. Her dissertation, "Crude Stages of the Anthropocene: Performance and Petroimperialism," examines a range of performance practices on oil frontiers, including civic festivals, political demonstrations, and political theater. She has contributed to *Theatre Journal, TDR: The Drama Review, Room One Thousand,* and the edited volume *Shakespeare beyond English: A Global Experiment.*

Kim Stanley Robinson is the author of nineteen science fiction novels, including the Mars trilogy.

Matthew Schneider-Mayerson is assistant professor of environmental studies at Yale–NUS College. He has published on literature, environmental politics, environmental futures, and popular culture. The author of *Peak Oil: Apocalyptic Environmentalism and Libertarian Political Culture,* he is currently writing a monograph on reproductive choices in the age of climate change and is editing books on empirical ecocriticism and the environmental dimensions of life in Singapore.

Ann Kristin Schorre holds a master's degree in human geography from the University of Oslo, Norway. Her research focuses on social change in agrarian settings with an emphasis on the role of culture in climate change adaptation. She has been working with the Adaptation-CONNECTS project led by Karen O'Brien.

Kate Shaw holds a bachelor of fine arts with honors from RMIT University and a diploma of museum studies from Deakin University. Her solo exhibitions include *Radiant Orb* (Mirus Gallery, San Francisco, 2016), ... (Fehily Contemporary, Melbourne, 2016),

Blue Marble (Turner Galleries, Perth, 2015), *Eternal Surge* (PointB, New York, 2014), *Stardust in Our Veins* (Fehily Contemporary, Melbourne, 2014), *ART14 London* (Fehily Contemporary, London, 2014), *Uncanny Valleys* (Cat Street Gallery, Hong Kong, 2014), *Luminous Worlds* (Gippsland Art Gallery, Victoria, 2014), and *Fjallkonan* (Fehily Contemporary, Melbourne, 2013). Her art has also been featured internationally in group exhibitions. She was a Wynne Prize finalist for several years, a Scegg Redlands Prize finalist, an Arthur Guy Memorial Painting Prize finalist, and a Substation Contemporary Art Prize finalist. She lives and works in Melbourne and New York.

Malcolm Sen is assistant professor of English at the University of Massachusetts Amherst. He is coeditor of *Postcolonial Studies and the Challenges for the New Millennium.* His podcast series, *Irish Studies and the Environmental Humanities*, is available through Google Play and Apple iTunes.

Sam Solnick is lecturer in English at the University of Liverpool, where he is codirector of the Literature and Science Hub. He is author of *Poetry and the Anthropocene.*

Caledonia Curry, who exhibits her art under the name SWOON, is a classically trained visual artist and printmaker who explores the relationship between people and their built environments. Her work has been collected and shown internationally at galleries and museums, including the Museum of Modern Art, New York; the Brooklyn Museum; the Institute of Contemporary Art, Boston; and the São Paulo Museum of Art. More information can be found at swoonstudio.org.

Rirkrit Tiravanija was born in Buenos Aires, Argentina. His art defies media-based description, as his practice combines traditional object making, public and private performances, teaching, and other forms of public service and social action. He is on the faculty of the School of the Arts at Columbia University and is a founding member and curator of Utopia Station, a collective project of artists, art historians, and curators. He helped establish The Land Foundation, an educational–ecological project in Chiang Mai, Thailand.

Miriam Tola is assistant teaching professor at Northeastern University. She specializes in feminist and postcolonial environmental humanities and critical theory. Her writing on the Anthropocene, the politics of the commons, and the rights of nature have been published in *Theory and Event, PhaenEx, South Atlantic Quarterly, Feminist Review,* and *Environmental Humanities.* She has worked as a journalist and documentary producer in Italy and the United States.

Sheena Wilson is professor at the University of Alberta, codirector of the international Petrocultures Research Group (petrocultures.com/), and lead researcher on *Just Powers* (justpowers.ca/). Her publications include *Petrocultures: Oil, Politics, Cultures;* "Gendering Oil: Tracing Western Petrosexual Relations"; and a coedited special issue of *Imaginations,* "Sighting Oil." She is writing the monograph *Deep Energy Literacy: Toward Just Futures.*

Daniel Worden teaches in the School of Individualized Study at the Rochester Institute of Technology. He is the author of *Masculine Style: The American West and Literary Modernism,* editor of *The Comics of Joe Sacco: Journalism in a Visual World,* and coeditor of *Postmodern/Postwar—and After: Rethinking American Literature* and *Oil Culture* (Minnesota, 2014).